The Green New Deal and Climate Change:

What You Need to Know

Lynne Balzer

FARADAY SCIENCE INSTITUTE
"A BETTER WORLD THROUGH SCIENCE AND TECHNOLOGY"

FaradayScience.org

ISBN #978-1-7334603-0-9

These chapters are dedicated to all those skilled and tireless individuals who, through their quest for the truth, have paved the way for knowledge and understanding about climate.

My special thanks go to Dr. Tim Ball and Dr. Don Easterbrook, who provided guidance and valuable insights in the creation of this project. I'm also grateful to Dr. Easterbrook and Tony Heller of RealClimateScience.com for sharing some of their valuable graphs.

My sister, Ellen, also deserves a special thanks for the encouragement she provided. Other members of my family: my son, Mark, who gave superb suggestions, grandson, Anthony, who provided valuable technical assistance and my daughter, Wendy, who patiently read through parts of the rough drafts.

The Green New Deal and Climate Change: What You Need to Know challenges the 50-year-old climate change story we've been given with facts that have been documented. For example,

- Since the dawn of Earth's history its climate has been changing - sometimes dramatically - without help from humans,

- For the past 20 years Earth's average temperature has **not** increased,

- Climate change alarmism began about 50 years ago with dire warnings about **global cooling,**

- The main "greenhouse gas" is water vapor, which is 70 times as potent as carbon dioxide, a **trace gas**.

- Humans contribute only about 4% of the carbon dioxide that is being released into the atmosphere,

- This small amount of extra carbon dioxide is greening the earth and increasing agricultural yields,

- The level of the oceans is rising at the same rate as it has for the past 1000 years: about **7" per century,**

- During the past fifty years the polar bear population has increased, not decreased.

- Climate change alarmism is **not** based on real science.

The science of climate is carefully explained by a respected science teacher in concrete, easy-to-understand terms.

This book is unique in that it also carefully analyzes the Green New Deal's proposed alternatives to fossil fuels, compared with energy solutions that are sustainable.

Table of Contents

Introduction

Because you are reading this book, you may already suspect that something is wrong with the idea of the "Green New Deal". But you may still be unaware of the reasons why it is wrong and how this all came about. You need the valuable information in this book both for your own peace of mind that you're on the right track and also to enable you to make informed voting/policy decisions and communicate accurately and effectively about this very important issue.

What if you were to wake up some morning to learn that everything you thought you knew was actually wrong? That would be a very disconcerting feeling indeed. You might think that you were in the Twilight Zone, or in a parallel universe. Of course this is an extreme case. Much of what you know is true. You may suspect that some of the "news" we hear each day is wrong. But most people aren't aware that a lot of what we see and hear on TV, in magazines, textbooks, newspapers and even Hollywood movies, is actually just a series of carefully crafted stories behind an agenda those in power have in mind for us. And it is being presented in a very convincing way: lies entwined with facts. The challenge is untangling the facts from the lies. Nowhere is this more evident than what the media has been saying about climate change.

Unfortunately we are living in a "post truth world" in which many people accept such claims without question. Too many of us base our decisions on beliefs and emotions rather than on facts and logic. Some people can't even distinguish between facts and opinions. Our schools must accept much of the blame for this. Many schools have become indoctrination centers that teach young people ***what*** to think rather than ***how*** to think. Emotions need to be replaced by critical thinking, which entails questioning everything we read and hear and demanding evidence to back up such claims. Ideas presented as facts must be checked against reality using logic rather than emotions.

But rather than investigating this issue on their own, many people rely on media they trust and on the scientists ("experts") promoted by these media trumpeting the dogma that humans are to blame for global warming. Most people don't have the science background to objectively analyze all of the conflicting information in order to make a valid judgment. A recent study revealed that only 20% of the world's population feel comfortable with science. In another study, conducted at Yale University, 50% of the students failed a test of climate knowledge.[1] However, this book explains the science you need to know to understand this issue.

Science is more than just knowledge about the environment. It's also a way of thinking and approaching problems. If the rules for how to do this are not followed, the science is flawed, and the conclusions reached are probably erroneous. Another important thing to remember is that science isn't omnipotent; it has its limitations. And scientists are human beings with the same frailties and human nature as everyone else.

Unfortunately, the barrage of propaganda to which the world's people have been subjected for the past four decades has turned this issue into a cause - even a religion - for many people. As evidence of religious climate hysteria, recently a woman working with an organization called Extinction Rebellion glued her breasts to the sidewalk in front of the Goldman Sachs building in London. At the London Stock Exchange thirteen people from the same organization glued themselves to one another and to a wall. These people were wearing signs stating, "climate emergency". An 83-year-old man climbed atop a train car from which he held a prayer vigil.[2]

A wise philosopher once told his students, "I reserve the right to change my mind about an issue when new information sheds light on it." The Internet has given us a unique and priceless opportunity to get information that past generations lacked. This library of the world enables us to check claims from different perspectives. We may not always have this opportunity, since powerful forces are already closing some doors to the truth by blocking or removing material that they don't want us to see. We therefore have no time to waste in taking advantage of the chance we still have to learn the truth.

---

1 Dr. Tim Ball, *Human-Caused Global Warming: the Biggest Deception in History*, Tellwell Talent, 2016, p. 40.

2 Lara Keay and Mark Duell, https://www.dailymail.co. uk/news/article-6958057/Extinction-Rebellion-protestors-glueLondon-Stock-Exchange-final-day-protests. html, April 25, 2019

This book carefully lays out the facts concerning the claims about impending doom caused by human-caused [anthropogenic] global warming (AGW) - or, as they say, "climate change" - and, just as important, the feasibility of their proposed solutions.

1. Why Everyone Needs to Understand the AGW[3] Issue

On January 21, 2019, freshman congresswoman Alexandria Ocasio-Cortez (D-NY) declared, "The world is going to end in 12 years [2030] if we don't address climate change."[4] She equated this problem with a war for our survival and its solution on the same level as the action the US took to help win World War II. According to her we are in a war against climate change.

Shortly thereafter, in a joint press conference, Ocasio-Cortez and Senator Ed Markey (D-MA) announced their Green New Deal, a proposal for a ten-year program that would revamp the US government, taking aggressive measures to deal with what its authors[5] perceive as the impending destruction of Earth due to runaway global warming. Nearly a hundred members of Congress including most of the Democratic presidential hopefuls have signed onto this plan. The media has given it the highest accolades.

The drastic measures outlined in this plan (see Appendix 2) rest on the assumption that the earth has been heating up faster than ever before due to mankind's emission of carbon dioxide (CO_2) into the atmosphere since the beginning of the Industrial Revolution. Carbon dioxide is produced by the combustion of all carbon based fuels, such as gasoline, wood, natural gas and coal. In addition, virtually all living organisms, even plants, produce carbon dioxide during respiration.

The United Nations Intergovernmental Panel on Climate Change (IPCC) report, SR15, has announced that humans are responsible for this warming and that an additional increase of 1.5°C in Earth's average temperature will likely bring about the extinction of many species of plants and animals and cause extreme weather events and danger to food supplies, human

---

3 AGW is the acronym for anthropogenic (human-caused) global warming

4 Christina Zhao, "Alexandria Ocasio-Cortez Warns 'World is Going to End In 12 Years,' Reiterating Claims of Recent U.N. Climate Change Report," *Newsweek*, Jan. 22, 2019

5 apparently unnamed

health and economic growth. This United Nations agency warns that if that problem isn't remedied within twelve years, it may be too late to save our planet from destruction. The IPCC estimate of the cost to do this is $2.4 trillion per year.[6]

However, as more information has surfaced over the years, many prominent scientists - even some Nobel prize winners - have been expressing serious doubts about the science behind this claim and whether there actually is a crisis. Their number is growing each year. These scientists have been largely ignored by government officials and the mainstream media. People who disagree with the "official view" have been called "climate deniers" or even "science deniers". Even worse, some politicians have called for jailing scientists who disagree with the IPCC's conclusion under RICO[7] laws. These scientists don't question the reality of climate change. Everyone knows that the earth's climate is and always has been changing. What they *are* questioning is whether *humans* are causing climate change and whether it's a serious problem.

The Green New Deal proposal calls for completely replacing, within this ten-year period, the burning of "fossil fuels" with renewable sources, such as wind, solar and biofuels. They want to replace all gasoline and diesel engines with electrically-powered cars, trucks, farm tractors and construction vehicles. Electric utilities would all rely on wind, solar and biofuel-powered generators. Although the combustion of wood, a biofuel, produces about 1.5 times as much CO_2 per unit of energy as does the combustion of coal, it's still considered a carbon-neutral, renewable source of energy by the European Union Renewable Energy Directive (RED).[8]

Backers of the Green New Deal also want to do away with nuclear power, even though in 2017 nuclear energy was producing 20% of the United States total electric energy output without emitting any CO_2.[9]

6 F. William Engdahl, "The Dark Story Behind Global Warming aka Climate Change", Oct. 16, 2018, journalneo.org

7 Racketeer Influenced and Corrupt Organizations Act: a federal law providing for criminal penalties for acts performed as part of an ongoing criminal organization

8 Timothy E. Searchinger, et al.,"Europe's renewable energy directive poised to harm global forests," Nature Communications volume 9, Article number: 3741 (2018), Sept. 12, 2018

9 https://www.world-nuclear.org/information-library/country-profiles/countries-t-z/usa-nuclear-power.aspx

If the Green New Deal goes into effect, your life will never be the same again. Financing and implementing this plan would involve a complete reshuffling of the US government and society. Your taxes would skyrocket. You'd be told what kinds of food to eat. Meat eating wouldn't be permitted, because the plan's proponents claim that the production of meat uses too many resources, which they say are limited. The raising of cattle and other farm animals for meat would be prohibited. Unless you own an electric car, you'll be traveling on railroad lines, walking or riding your bike. According to this plan airline travel would also be phased out.

To help pay for this plan the Federal Reserve Bank would be called upon to print and put into circulation trillions of dollars in paper money. Putting more paper money into circulation would create serious inflation, devastating everyone's savings and retirement accounts. We need only to look at what recently happened in Venezuela. The rate of inflation that just occurred there under a socialist dictatorship was as high as 1.6 million percent.

Under the Green New Deal every building in the United States would either be re-engineered or rebuilt for energy efficiency. Every aspect of your life would be micromanaged by unelected bureaucrats. This is true socialism at best, but most likely, totalitarian communism, even though those terms are omitted from the document.

Before allowing our lives to be taken over in an experiment to "prevent the world from ending", we all need to learn the actual facts behind the global warming scare. Is the earth really headed for disaster if we fail to go through with this plan or a similar action?

Have our elected representatives who are backing this plan considered the consequences of such action? Do they think we can save polar bears from extinction by changing to a different type of light bulb? Or that we can stop ocean levels from rising by building wind turbines? Or make hurricanes less severe by driving electric cars? Are the thousands of news reports to which we've been subjected during the last 37 years really accurate? Have we been told the truth?

The three main tenets of the belief in human-caused [anthropogenic] global warming are:

1) Carbon dioxide [CO_2] is the principal greenhouse gas.

2) Carbon dioxide controls global temperature, which has been steadily increasing.

3) Man can control the climate.

Our governments and the media have informed us of the impacts of climate change, which they say have already occurred. These include:

- more heat waves but also more cold spells,
- more droughts but also more floods,
- less snow but also more snow,
- stunted agricultural production but also expanded agricultural production,
- more mental illness and suicides,
- more alcoholism,
- more crime,
- more bugs, rats and bats,
- the civil war in Syria,
- more earthquakes,
- more tsunamis,
- more icebergs but also fewer icebergs.[10]

10 Lecture by Prof. Murry Salby at the Sydney Institute, published on Feb. 26, 2019

This issue really boils down to energy. The production of energy that humans desperately need for transportation, heating, water and the production of goods and services makes all the difference between the living standards we now enjoy and those of not too long ago, when energy and power were provided by slaves and beasts of burden.

The first part of this book examines from a scientific standpoint the claim that is the basis of the Green New Deal, cap and trade and various other carbon tax plans: that Earth will shortly be destroyed by runaway global warming if we don't drastically reduce our CO_2 emissions - our "carbon footprint". The second part compares the alternatives to the combustion of carbon based fuels. As a voter or other decision maker, you need to know as much as possible about the entire situation. This way you'll know which candidates for office to support: those who want to take drastic action to reduce CO_2 or those who believe that global warming is not a crisis.

If you lack the confidence to make an informed decision about this important issue because you feel that you have inadequate scientific knowledge or qualifications, this book will provide a clear explanation that any reasonably educated person can understand. You don't need a PhD in climate science to understand the problem and its solution once it's been presented in concrete down-to-earth terms.

2. Failed Climate Predictions

Philip Stott aptly pointed out, "In essence, the Earth has been given a 10-year survival warning regularly for the last fifty or so years...Our post-modern period of climate change angst can probably be traced back to the late-1960s...By 1973, and the 'global cooling' scare, it was in full swing, with predictions of the imminent collapse of the world within ten to twenty years...Environmentalists were warning that, by the year 2000, the population of the US would have fallen to only 22 million. In 1987, the scare abruptly changed to 'global warming'".[11]

The early twentieth century had experienced a period of warming, and the decade of the 1930's to 1940's was the warmest in recent experience. By the 1960's and 1970's, it was evident that temperatures had started to decrease even though atmospheric CO_2 was rising. If CO_2 were the culprit, we have to wonder how the earth could have been cooling down at the same time.

On Earth Day 1970 Dr. Kenneth Watt warned, "The world has been chilling sharply for about twenty years. If present trends continue, the world will be about four degrees colder for the global mean temperature **in 1990,** but eleven degrees colder **in the year 2000**. This is about twice what it would take to put us into an ice age."

On July 9, 1971, an article in the *Washington Post* warned, "**In the next 50 years** [2021] fine dust that humans discharge into the atmosphere by burning fossil fuel will screen out so much of the sun's rays that the Earth's average temperature could fall by six degrees. Sustained emissions over five to ten years, could be sufficient to trigger an ice age."[12]

An article in the April 28, 1975, issue of *Newsweek Magazine* declared, "There are ominous signs that Earth's weather patterns have begun to change dramatically....The evidence in support of these predictions has now begun to accumulate so

11 Engdahl, op. cit.
12 www.aim.org/aim-column/washington-post-gives-megaphone-to-far-left-groups-

massively that meteorologists are hard-pressed to keep up with it....The central fact is that...the earth's climate seems to be cooling down...If the climate change is as profound as some of the pessimists fear, the resulting famines could be catastrophic."

To this, in June of 1975, Nigel Calder in *International Wildlife,* added, "The continued rapid cooling of the earth since WWII is in accord with the increase in global air pollution associated with industrialization, mechanization, urbanization and exploding population."

In the 1970's carbon dioxide, a product of combustion, as well as human-caused air pollution in general was blamed for the fact that Earth was *cooling down*! Scientists blamed dust produced by mankind's activities that was circulating around Earth and blocking solar radiation.

Articles about global cooling abounded in magazines, newspapers, TV reports and books. Some of the suggested solutions to this "crisis" included taking away our rights and freedoms. Arnold Reitze, a professor at Case Western Reserve University warned,

> "We will be forced to sacrifice democracy by the laws that will protect us from further pollution." He suggested outlawing internal combustion engines, placing rigid controls on all research and development in the production and marketing of new products and population controls, with a penalty imposed on couples producing more than the prescribed number of children."[13]

This idea that overpopulation was the problem coincided with Paul Ehrlich's 1968 book, *The Population Bomb.* In asserting that the world's rapid population growth was unsustainable, Ehrlich proposed radical population control measures — including sterilization. In 1977 Ehrlich and John Holdren, who later became Obama's science czar, together wrote a book entitled, E*coscience: Population Resources, Environment.* This book called for forced abortions, contraception drugs in water supplies, seizing babies from single and teen mothers, forced

---

13 "Cold Facts About Pollution," *The Southeast Missourian*, Jan. 29, 1970

sterilization of those they deemed unfit and an international police force.[14] This concept of eugenics, which inspired the Holocaust perpetrated by the Nazi's, has been around for a long time.

In his 1976 book, *The Cooling,* Lowell Ponte declared, "This cooling has already killed hundreds of thousands of people. If it continues and no strong action is taken, it will cause world famine, world chaos and world war, and this could all come about **before the year 2000**." Today's global warming alarmists are hoping that nobody remembers these dire predictions, but they are on record.

The preface of Ponte's book reads,

> "It is a cold fact: the global cooling presents humankind with the most important social, political, and adaptive challenge we have had to deal with for ten thousand years. Your stake in the decisions we make concerning it is of ultimate importance; the survival of ourselves, our children, our species."

When the temperatures started to rise again around 1980, the alarmists changed gears and - all in step - started issuing *opposite* predictions, with anthropogenic [human caused] carbon dioxide still being the culprit.

In 1982 Mostafa Tolba, executive director of the UN Environment Program (UNEP), stated that the "world faces an ecological disaster as final as nuclear war within a couple of decades unless governments act now." His prediction was that if the world failed to act, there would be "by the turn of the century, an environmental catastrophe which will witness devastation as complete, as irreversible as any nuclear holocaust."[15]

In a 1988 interview Dr. James Hansen of NASA's Goddard Institute for Space Studies (GISS) was asked how he thought the the area around his office in New York would be affected in **20 years**, [or 2008]. Hansen replied:

---

14 Dr. Tim Ball, *Human Caused Global Warming: The Biggest Deception in History,* Tellwell Talent, 2016, p. 85.

15 Engdahl, op. cit.

> "The West Side Highway [which runs along the Hudson River] will be under water. And there will be tape across the windows across the street because of high winds. And the same birds won't be there. The trees in the median strip will change....There will be more police cars....[since] you know what happens to crime when the heat goes up."[16]

These days Dr. Hansen is still issuing equally alarming predictions. In a 2016 paper published in *Atmospheric Chemistry and Physics,* the global-warming guru stated:

> "The current rate of global warming could raise sea levels by 'several meters'[17] over the coming century, rendering most of the world's coastal cities uninhabitable and helping unleash devastating storms...Global warming of 2°C above pre-industrial times – the world is already halfway to this mark – would be "dangerous" and risk submerging cities."[18]

Recently Hansen also remarked that anyone who disagrees with him should be jailed.

However, although sea level has been rising at a constant rate since the last ice age ended 10,000 years ago, it has been only about seven inches per century (1.5mm/yr), and reportedly its rate of increase is slowing down.[19]

On June 30, 1989, the Associated Press released the following misinformation:

> "U.N. Official Predicts Disaster, Says Greenhouse Effect Could Wipe Some Nations Off Map. Entire nations could be wiped off the face of the earth by rising sea levels if global warming is not reversed **by the year 2000**. Coastal flooding and crop failures would create an exodus of 'eco-refugees,' threatening

16 https://stevengoddard.wordpress.com/2013/03/10/another-hansen-fail/
17 "Several" could be five or more meters. Several meters equals about 17 feet, or from the ground to the roof of a house.
18 www.theguardian.com/science/2016/mar/22/sea-level-rise-james-hansen-climate-change-scientist
19 https://www.scientificamerican.com/article/a-scientist-explains-the-mystery-of-recent-sea-level-drop/

> political chaos."

Brown, who was the director of the New York office of the U.N. Environment Program, added that governments had a **10-year** window of opportunity **[1999]** to solve the greenhouse effect.[20]

Despite any propaganda you may have heard from the mainstream media, no countries have suffered any such damage, or been "wiped off the map".

A Sept 19, 1989, article in the *St. Louis Post-Dispatch* stated, "New York will probably be like Florida **15 years from now.**" That would have been 2004.

December 5, 1989, in the *Dallas Morning News,*

> "Some predictions for the ***next decade*** [1989-1999] are not difficult to make...Americans may see the '80s migration to the Sun Belt reverse as a global warming trend rekindles interest in cooler climates."

In 1990 Michael Oppenheimer of The Environmental Defense Fund made the following ominous prediction:

> "**By 1995**, the greenhouse effect would be desolating the heartlands of North America and Eurasia with horrific drought, causing crop failures and food riots..."**(By 1996)** "The Platte River of Nebraska would be dry, while a continent-wide black blizzard of prairie topsoil will stop traffic on interstates, strip paint from houses and shut down computers...The Mexican police will round up illegal American migrants surging into Mexico seeking work as field hands."[21]

None of this has happened, of course.

On April 22, 1990, it was reported on ABC's *The Miracle Planet,*

---

20 www.theintermountain.com/opinion/letters.../disaster-predictions-often-turn-out-false/

21 http://www.foxnews.com/tech/2010/12/30/botched-environmental-forecasts.html

> "I think we're in trouble. When you realize how little time we have left - we are now given not **10 years** to save the rain forests, but in many cases **five years**. Madagascar will largely be gone in **five years** [1995] unless something happens. And nothing is happening."[22]

Madagascar, located in the Indian Ocean off the eastern coast of Africa, is the fourth largest island in the world. It has not been affected by the above mentioned catastrophes.

On October 15, 1990, renowned scientist Carl Sagan remarked, "The planet could face an 'ecological and agricultural catastrophe' **by the next decade** [2000] if global warming trends continue."[23]

In his 1991 book entitled, *5000 Days to Save the Planet,* Edward Goldsmith wrote,

> "By 2000, British and American oil will have diminished to a trickle....Ozone depletion and global warming threaten food shortages, but the wealthy North will enjoy a temporary reprieve by buying up the produce of the South. Unrest among the hungry and the ensuing political instability, will be contained by the North's greater military might. A bleak future indeed, but an inevitable one unless we change the way we live...At present rates of exploitation there may be no rainforest left in 10 years. If measures are not taken immediately, the greenhouse effect may be unstoppable in **12 to 15 years**."

This deadline would have been between 2003 and 2006.

February 1993, Thomas E. Lovejoy of the Smithsonian Institution remarked, "Most of the great environmental struggles will be either won or lost **in the 1990s,** and **by the next century** [2001] it will be too late."[24]

---

22 http://anodtothegods.com/?p=9061

23 https://realclimatescience.com/2017/10/hot-off-the-press-todays-climate-fraud-from-nrdc/

24 https://www.nytimes.com/1989/02/28/opinion/in-the-nation-decade-of-decision.html

This is but one of many "tipping point" statements in which we were told that if people didn't drastically change the amount of CO_2 they were putting into the atmosphere, it would be "too late".

On November 7, 1997, a BBC commentator remarked,

> "It appears that we have a very good case for suggesting that the El Niños are going to become more frequent, and they're going to become more intense and **in a few years, or a decade or so**, we'll go into a permanent El Niño. So instead of having cool water periods for a year or two, we'll have El Niño upon El Niño, and that will become the norm. And you'll have an El Niño, that instead of lasting 18 months, lasts 18 years."[25]

"In a few years" would have been by the year 2000. This just didn't happen.

July 26, 1999 *The Birmingham Post* (UK),

> "Scientists are warning that some of the Himalayan glaciers could vanish **within ten years** [2009] because of global warming. A build-up of greenhouse gases is blamed for the meltdown, which could lead to drought and flooding in the region affecting millions of people."[26]

This would have occurred by 2009. Later the the United Nations Intergovernmental Panel for Climate Change (IPCC) made this retraction: "The UN's climate science body has admitted that a claim made in its 2007 report - that Himalayan glaciers could melt away by 2035 - was unfounded.

"The admission today followed a *New Scientist* article last week that revealed the source of the claim made in the 2007 report by the Intergovernmental Panel on Climate Change (IPCC) was not peer-reviewed scientific literature - but a media interview with a scientist conducted in 1999."[27]

25 news.bbc.co.uk/2/hi/science/nature/25433.stm
26 misesuk.org/2014/04/03/watts-big-list-of-failed-global-warming-predictions/
27 /www.theguardian.com/environment/2010/jan/20/ipcc-himalayan-glaciers-

March 20, 2000, in *The Independent*, According to Dr. David Viner of the University of East Anglia Climatic Research Unit, snowfall in Britain would become "a very rare and exciting event" and "Children just aren't going to know what snow is."[28] The fact is that on December 10, 2017, *The Telegraph* reported:

> "The worst snowstorms in Britain for four years closed roads, left whole towns without power and grounded flights.
>
> "The Army was placed on standby and the police urged motorists not to travel unless their journeys were absolutely necessary. Persistent heavy snowfall hit the Midlands, the home counties and Wales throughout the morning, leading the Met Office to issue an amber weather warning...Flights were suspended at Birmingham and Luton airports while the runway was cleared, leaving passengers facing lengthy delays. Flights into Heathrow were diverted to Newcastle, 300 miles away."[29]

This was just the first of many snowstorms experienced in Great Britain and all over Europe during the winter of 2017-2018, one of the coldest and snowiest in recent history.

Global-warming alarmists claim that it was cold in the rest of the world due to the abnormally warm weather at the North Pole. One problem with this claim is that there is reportedly only one real weather station in the entire Canadian Arctic region, which prevents accurate record keeping. The fact is that these temperatures are simply estimated, not scientifically recorded.

In January of 2000 Dr. Michael Oppenheimer of the Environmental Defense Fund, commenting in a NY Times interview on the mild winters in New York City, exclaimed:

> "But it does not take a scientist to size up the effects of snowless winters on the children too young to

mistake

28 https://wattsupwiththat.com/2010/12/24/snowfall-a-very-rare-and-exciting-event/

29 www.telegraph.co.uk/news/2017/12/10/roads-closed-flights-grounded-snow-spreads-across-uk/

> remember the record-setting blizzards of 1996. For them, the pleasures of sledding and snowball fights are as out-of-date as hoop-rolling, and the delight of a snow day off from school is unknown."[30]

However, this just didn't happen. Most recently New York City experienced several significant snowfalls during the winter of 2017-2018. On March 12, 2018, a blizzard with gale winds up to 50 mph hit eastern North Carolina, several hundred miles south of New York. Parts of Kentucky received about a foot of snow in the same system.

On March 29, 2001, CNN reported: "In **ten year's time** [2011], most of the low-lying atolls surrounding Tuvalu's nine islands in the South Pacific Ocean will be submerged under water as global warming rises sea levels."

Yet on Feb. 9, 2018, new research reported in an article in *the Daily Mail* shows that the land area of this island, which is comprised of a coral atoll, is actually growing in size. This study conducted by the University of Auckland in New Zealand used aerial photographs and satellite imagery to demonstrate that eight of the atolls and nearly three fourths of the islands grew in size between 1971 and 2014.[31]

In 2005, Andrew Simms, policy director of the New Economics Foundation, announced, "Scholars are predicting that 50 million people worldwide will be displaced **by 2010** because of rising sea levels, desertification, dried up aquifers, weather-induced flooding and other serious environmental changes." This simply didn't happen.

June, 2008, Ted Alvarez, *Backpacker Magazine Blogs*: "You could potentially sail, kayak, or even swim to the North Pole by the end of the summer. Climate scientists say that the Arctic ice...is currently on track to melt **sometime in 2008**."[32] Later in that same year a Russian icebreaker was stranded for a week

---

30 http://www.climatedepot.com/2014/02/18/flashback-2000-new-york-times-sledding-and-snowball-fights-are-as-out-of-date-as-hoop-rolling/

31 http://www.dailymail.co.uk/sciencetech/article-5371535/Sinking-Pacific-nation-getting-bigger-study.html

32 https://thebigfeedblog.blogspot.com/2008/12/the-worst-climate-predictions-of-2008.html

in the ice of the Northwest Passage.[33] Arctic ice has been increasing.

As the projected dates for these dire predictions that were supposedly based on science came and went during the past four decades and none of them materialized, it became evident that they were based on nonsense.

33 http://losingamerica.truth-in-politics.info/?p=3142

3. The "Science" Behind the AGW Hypothesis

A hypothesis is an idea that has not yet been proven. The AGW hypothesis states not only that there has been global warming but also that *mankind* is responsible for the warming.

Many younger people are unaware of the length of time the global cooling and global warming propaganda scares have been going on. Yet, as explained before, *not a single one* of the predictions made by these alarmists has materialized. The crisis they say we are facing has three postulates:

1. Mankind's activities are responsible for runaway global warming.

2. Industrialized nations especially must atone, or pay the price, for this warming.

3. Man is able to control the climate.

Predictions of the amount of future warming on which the IPCC bases its claims comes from about 100 computer models. The problem here is that if you don't correctly program into the computer *everything,* including the tiniest detail, that is occurring in the atmosphere, you won't get a valid result. The earth's weather patterns are very complicated. The world's top climatologists admit that they are still not well understood. This is one reason why the predictions of the models have been so far off base, predicting warming of about three to four times that which actually has occurred in the last 39 years.

To reduce the amount of error, the people putting information into these models have periodically tweaked, or tuned them to conform to *past* records. In spite of this, they still err by a great deal, as you can see on the following graph.

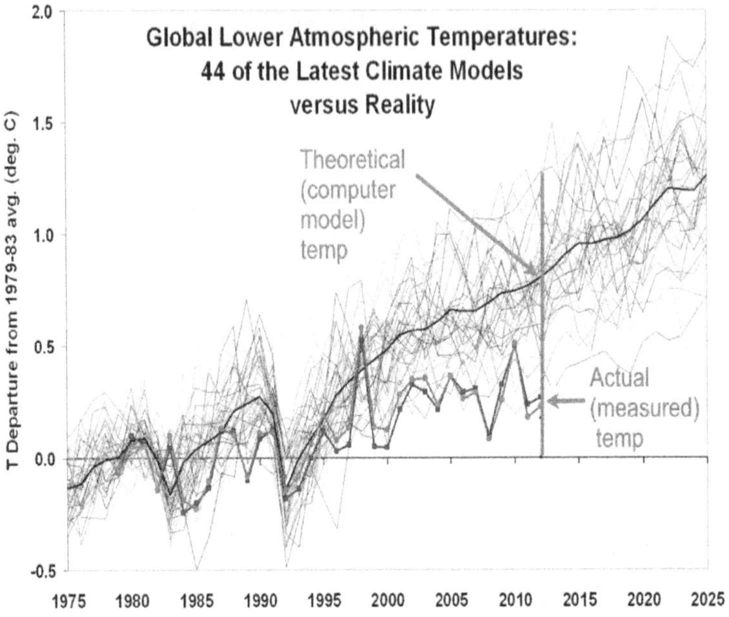

There is a huge difference between the observed temperature data and the predictions graphed above, especially starting about 2000. True science is empirical. That means that it is based on observations, using our five senses, and measurements of the *real world*. In order to study weather and climate - which is weather over an extended period of time - temperature, rainfall, wind speed and direction, air pressure and sea ice extent are carefully measured, monitored, and recorded.

The pieces of information gained in these measurements and tests are called data (a plural word). Data are carefully analyzed and used to draw conclusions and make predictions. This procedure is known as the scientific method. If the results or predictions of a hypothesis (initial idea) turn out to be incorrect, then it follows that either the hypothesis or the methods used to test it are wrong. Scientists then must figure out why they didn't get the result they expected. The hypothesis might have to be scrapped or changed.

After a report of such a scientific study is carefully written out by the researchers, it must then be submitted to a scientific journal for publication. This important process gives other scientists in that field (called peers) the opportunity to read, evaluate and question the data obtained in the study and the conclusions reached. This also enables them to repeat the investigation, which should yield a similar result. This important step of the scientific method is all part of what is called peer review.

Because observations and experiments must be repeated over and over again, scientific investigation is difficult, time consuming and often boring. It could take years to complete a study and more years for other teams working in the same field to verify the results. Only then can a hypothesis be considered a "theory". Even established theories are questioned and sometimes overturned. It is the very nature of science to be skeptical. While models can be helpful at times, they cannot pass for empirical science. Nor will they ever be able to replace empirical science.

Climate alarmists attribute the 0.8°C. rise in global average temperature since the mid 1800's to the increase in the percentage of atmospheric carbon dioxide from 0.032% to ~0.04% at present. They claim that human emitted carbon dioxide, added to natural sources of this gas, is the culprit.

When you get into your car on a sunny day, you observe that the temperature inside is much warmer than it is outside. Just as in a greenhouse, sunlight entering through the glass is trapped inside as heat.

Greenhouses and automobiles, having walls and roofs, are closed systems. Earth is not a closed system, however. Its atmosphere has no walls or roof as does a greenhouse. *All* of the sun's radiation falling on Earth's surface is radiated back into space. If this weren't true, Earth would just keep getting warmer and warmer.

A few atmospheric gases, such as water vapor, carbon dioxide and methane, hold onto some of this incoming energy for a short time before releasing it. These are called greenhouse gases. This so-called "greenhouse effect" enables life on Earth to exist. Without this phenomenon the earth's average temperature, which is about 15°C, would instead be -18°C, 33 degrees colder, and life as we know it couldn't exist. But a greenhouse effect that was out of control could theoretically raise temperatures higher than we'd want them to be.

So what is the role of carbon dioxide, a "greenhouse gas," in the earth's climate? Contrary to what you've probably heard, the main greenhouse gas is not CO_2 at all but water vapor. Comprising about 3% of Earth's atmosphere, water vapor's role in the greenhouse effect is at least seventy times as responsible for warming as is carbon dioxide, which makes up only about 0.04% of the atmosphere: four hundredths of one percent, or 400 parts per million (ppm).

The primary source of CO_2 in the atmosphere is its release from oceans, which cover about 70% of Earth's surface and from the earth's interior by volcanoes. The earth has over 1500 active volcanoes, many of which are spewing lava underwater through cracks where tectonic plates are separating. Much of this carbon dioxide was stored in the earth's mantle when our planet formed billions of years ago.

Carbon dioxide is also emitted by dead decaying organisms, bogs and swamps. The continuous process by which carbon is exchanged between plants, animals, rocks, the atmosphere and the oceans is known as the carbon cycle.

A study reported by the *New York Times* concluded that termites, digesting vegetable matter with the help of bacteria, produce more than twice as much carbon dioxide as all the world's smoke stacks.[34] Termite activity has been particularly high due to the widespread clearing of land, which has provided abundant food for them in the leftover debris. When termites, aided by bacteria, digest this organic matter, they add carbon dioxide and methane, another greenhouse gas, to the atmosphere.

It is a myth that when carbon dioxide is released, it moves up into the atmosphere. This is scientifically impossible. Because carbon dioxide is more dense than oxygen and nitrogen, the atmosphere's two main gases, it instead sinks downward, where it nourishes plants and seeps into the soil and bodies of water,

34 https://www.nytimes.com/1982/10/31/us/termite-gas-exceeds-smokestack-pollution.html

joining with other elements such as calcium to form animal shells and rocks. Contrary to popular myths, CO_2 doesn't remain in the atmosphere for long periods of time. A typical carbon dioxide molecule is in the atmosphere for only about fifteen days.[35]

Carbon dioxide *is essential to all life on earth*. This is because green plants and bacteria use it to make food for themselves. In the chemical process of photosynthesis carbon dioxide reacts with water in the presence of light to form sugars and starches that nourish plants and thus also keep animals alive.

It's been found that plants thrive best on a concentration of about 1200 ppm of CO_2 (0.12% of the atmosphere). This is three times the present level. For that reason this life giving gas is regularly pumped into commercial greenhouses, increasing yields by up to four times. This is empirical evidence, not speculation.

World renowned scientist, Dr. Will Happer, a specialist in adaptive optics, stated that at the present time the percentage of atmospheric carbon dioxide is unusually low, and plants are actually suffering from this shortage of CO_2.[36] If the amount of CO_2 in the atmosphere ever fell below 150 ppm, almost all living things would die. This fact is confirmed by Dr. Patrick Moore, past president of Greenpeace.

Carbon dioxide absorbs and emits energy only within a very limited frequency range. It briefly holds photons of light of just two wavelengths: 4.26 micrometers (asymmetric stretching mode) and 14.99 micrometers (bending vibrational mode). Once that small frequency range has been saturated, further additions of CO_2 to the atmosphere make virtually no difference. Dr. Happer equated this phenomenon with painting a barn red. Although the first couple of coats of paint will make a difference, subsequent coats have no effect on the color. The barn is already as red as it can be.

The influence of greenhouse gases is measured by atmospheric opacity, as opposed to oxygen and nitrogen, that are transparent to the passage of light. Changes in opacity

35 Dr. Fred Goldberg lecture, (YouTube video) March 15, 2016
36 William Happer, Professor of Physics at Princeton Unitersity, "Carbon Dioxide, Friend or Foe?", YouTube video, Nov. 10, 2010

decrease as further additions of CO_2 occur. Changing the ppm of CO_2 from 0 ppm to 280 ppm increases atmospheric opacity about 6.2%. Increasing it from 280 ppm to 400 ppm makes a difference of only 0.6%, one-tenth as much. Only about 4% of that increase comes from human-caused [anthropogenic] emissions, making a difference of only 0.2%.[37] Even if CO_2 should rise to 690 ppm in the 22nd century, the atmospheric opacity from CO_2 would increase by only another 1%, of which only a small amount would be caused by humans.

The earth has become noticeably greener as a result of the increase in carbon dioxide since the early 20th century. (Compared with a 100-yard-long football field, this increase in carbon dioxide would amount to three-eighths of an inch.) This small rise in CO_2 has been a boon to agricultural production, which has increased about 15% since then. This greening of Earth's surface has even been observed in satellite photos.

Yet in December of 2009 the Environmental Protection Agency (EPA) under the Obama administration classified this life giving compound as a pollutant.[38] People posing as environmentalists and the media have been using the phrase "carbon pollution" to describe the release of carbon dioxide into the atmosphere by humans.

Taking a close look at the graph below showing the temperatures and carbon dioxide levels of the past 600 million years, you can see that, currently at 400 ppm, CO_2 is at one of its lowest levels. In contrast, approximately 432 million years ago there was an Ice Age (Ordovician) in which the CO_2 level was 4500 ppm, over ten times as high as today's level. The only other time in geologic history that the CO_2 level was at its present level was about 250 million years ago.

For the last 250 million years the average level of carbon dioxide has been about 1200 ppm, the optimum level for plants. This graph shows that there is no correlation between temperature and CO_2. However, there ***is*** a correlation between CO_2 and plant growth.

---

37 Prof. Murry Salby, "Climate Scientist Murry Salby Demolishes the Global Warming Alarm," lecture on YouTube, Aug. 13, 2014

38 http://dailycaller.com/2015/02/11/epa-challenged-carbon-dioxide/

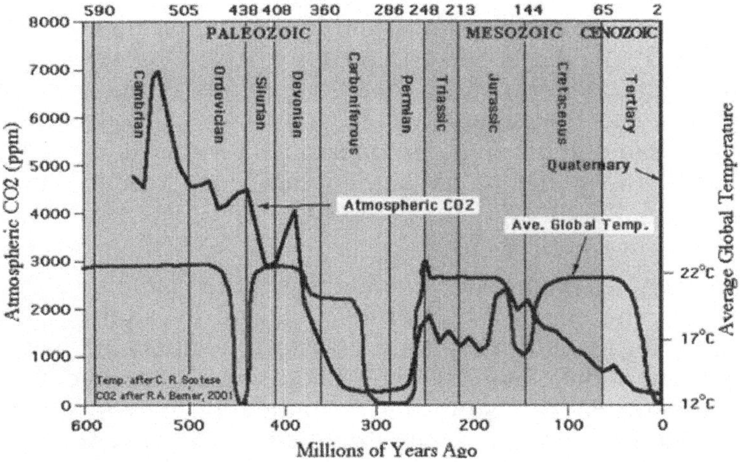

The photos of ominous looking clouds rising from the chimneys of coal burning plants, portrayed in the media, are actually clouds, *not* carbon dioxide. Whereas you can see clouds - made of tiny drops of *liquid water* - carbon dioxide is an *invisible* gas which doesn't show up in photographs. People are taken in by these misleading photos that are sometimes even touched up on Photoshop to increase contrast, making them appear menacing. This doesn't mean that smokestacks might not be emitting harmful chemicals such as sulfur dioxide. It's just that carbon dioxide, an invisible gas, isn't one of them.

This image is not of carbon dioxide but of clouds, which are made up of tiny drops of *liquid* water. In contrast, carbon dioxide is a colorless, odorless gas that is essential to life on Earth.

Some people have also confused life-giving carbon dioxide (CO_2) with carbon monoxide (CO), an extremely toxic gas. The difference between these two compounds is one oxygen atom per molecule. Because carbon monoxide has a greater affinity for the hemoglobin in a person's bloodstream than oxygen does, it takes the place of oxygen in the bloodstream. The person literally suffocates.

The computer climate modelers of the 1980's and beyond guessed that the slight warming caused by adding CO_2 to the atmosphere would cause more water to evaporate, and that this extra humidity would then hold in more heat, raising temperatures. This supposed amplification by water vapor was programmed into the computer models, causing them to make wild predictions of future warming. In the ensuing thirty nine years these predictions have been proven wrong by observational [empirical] data. Satellite temperature measurement shows that in the first 18 years of the 21st century there was no warming, and possibly even some cooling. By 1999 evidence that the predicted amplification by water vapor wasn't occurring had already come in. This fact was finally released in an obscure science journal in 2006.[39]

A team headed by Dr. Richard Lindzen of MIT conducted a study published in 2009 showing that when Earth receives more heat from the sun, it simply radiates more heat into space.[40] Another climatologist, Roy Spencer, reported in 2007 that during periods of warming there are actually fewer high clouds to hold in heat.[41] This acts as a buffer that counteracts any feedback from water vapor predicted in the 1980's. Low clouds reflect some of the sun's radiation, cooling the Earth.

It turns out that clouds are the key to understanding climate. Clouds can either hold in heat or block sunlight from reaching Earth's surface, cooling it. This depends on their height. High clouds generally slow down the dissipation of heat into space, causing warming.

On the other hand, low clouds block sunlight, which is reflected back into space. This is called the albedo effect. If more low

39 Topher Field, "50 to 1David Evans Interview," YouTube, Sept. 1, 2013
40 Knappenberger, https://www.masterresource.org/.../climate-sensitivity-estimates-heading-down-way-down
41 joannenova.com.au/2009/12/the-one-flaw-that-wipes-out-the-crisis/

clouds than high clouds form as a result of this extra humidity, Earth is cooled. It is a negative feedback which actually offsets the slight warming effect of CO_2.

Beginning in the 1980's computer models also predicted that if CO_2 was causing warming, there would be a "hot spot" over the tropics, called a global warming "signature". The idea was that the resulting warm water vapor in the tropics would rise to a higher level of the troposphere, about six miles up. For decades thousands of radiosonde balloons sent aloft in that area have failed to detect this hot spot. Therefore, many climatologists now feel quite certain that if that hot spot existed, it would have been detected by now.[42]

There are actually five different temperature datasets: two sets of satellite temperature records from the University of Alabama in Huntsville (UAH) and three land-based datasets: Hadcrut in England, NCDC and the NASA GISS temperature dataset. Anyone can download this data from the Internet, put it onto an Excel spreadsheet and run a trend line through the data.

Satellite data has proved to be more accurate than surface weather stations for several reasons. Thermometers used in ground-based weather stations are often located close to air conditioner vents spewing warm air as well as in airports and other paved areas where concrete heated by the sun raises the temperature. A century ago, most weather stations were located in grassy, unpaved, rural areas.[43] The influence of human activities such as traffic, concrete and air conditioner exhaust vents in cities is known as the urban heat island (UHI) effect. This can make a difference of up to several degrees in temperature measurement, skewing the results. In many cases cities grew up around weather stations that had previously been in rural areas. A team of volunteers headed by meteorologist Anthony Watts checked US weather stations to spot those with such problems.

Also, surface weather stations don't measure Earth's average temperature accurately, because they're not evenly distributed

---

42 David Evans, *The Australian,* July 18, 2008

43 David Evans, "Is the Western Climate Establishment Corrupt?", Nov. 8, 2010, http://scienceand publicpolicy.org/images/stories/papers/originals/western_climate_establishment _corrupt.pdf

around the globe. The majority of these surface weather stations are located in the United States. But comparatively few weather stations are located in the southern hemisphere, in the Arctic region or in oceans, which make up 70% of Earth's surface. Shockingly, in computing Earth's average temperature, many temperatures in those regions are simply guessed at. Land based weather stations cover only about 25% of the earth. So statements that we just had the "warmest year on record", when the average temperature has been recorded as a few hundredths of a degree higher than the previous estimate, are meaningless. This simply indicates that temperatures were flat.

Temperature records have been kept for only about 140 years. Other methods of estimating temperature, such as ice core samples, reveal the many climate changes which have occurred during Earth's climate history.[44]

In contrast to land based thermometers, satellite temperature sensing equipment measures all of the earth's surface equally. The ARGO system has been used since 2003 to measure ocean temperatures. "There are about 4,000 Argo floats in the ocean. Most of the time they are asleep, a thousand metres below the surface. Every 10 days they wake up and slowly rise to the surface, taking temperature measurements as they go. When they reach the surface, they radio their data back to headquarters, slip beneath the waves, sink down to a thousand metres and go back to sleep."[45]

This ocean temperature measurement system reveals a surprising amount of stability, with some areas warming about 2° per decade and some cooling at about -1.5° per decade. Overall, the ocean warmed just a slight amount, about 0.03°C per decade.[46]

Both the satellite and ARGO systems have determined that there were no measurable temperature increases either of the atmosphere or the oceans for the first nineteen years of the 21st century.

---

44 Ball, op. cit., p. 8
45 Willis Eschenbach, "Argo, Temperature, and OHC," https://wattsupwiththat.com/2014/03/02/argo-temperature-and-ohc/
46 Ibid.

4. Earth's Climate History

The percentage of atmospheric CO_2 has varied considerably throughout different periods of geologic history. Ice core samples show that there have been periods when the percentage of carbon dioxide was much higher at times when Earth was cooler than it is now. A graph of climate history plotting the amount of CO_2 with the average temperature shows that a rise or fall in temperature always occurs *before* the rise or fall in carbon dioxide levels. The average lag time is about eight hundred years. When ocean water warms following a warm period, some of the CO_2 dissolved in this water is released into the atmosphere. This is because more carbon dioxide can dissolve in cold water than in warm water. That's a scientific fact.

Proxy data (other than direct measurement by thermometers) taken from ice cores drilled in Antarctic and Greenland glaciers reveal the temperatures and percentage of various gases going back to around 900,000 years ago. Samples of ice cores obtained before 1999 lacked the precision to determine the correlation between CO_2 and temperature. Later improvements in ice core technology revealed that increases in temperature

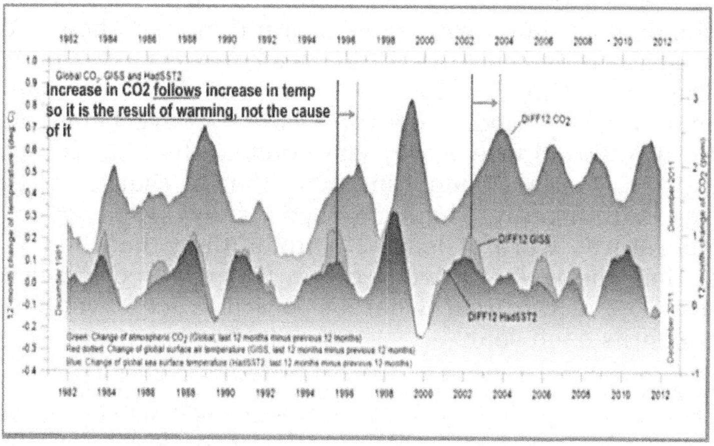

always precede increases in CO_2. This fact, known by all climatologists by the year 2003, shows that an increase in CO_2 is the *result,* not the *cause* of increased temperature. The average lag time is around 800 years, as shown on this graph. This is irrefutable proof that increases in carbon dioxide don't cause increases in temperature. It is just the opposite.

There are about 800 gigatons (billion tons) of carbon dioxide in the atmosphere. Human activities contribute only about 4% of this atmospheric carbon dioxide. But there is about fifty times as much CO_2 - about 40,000 gigatons - dissolved in the oceans.[47]

During its history Earth experienced several ice ages during which glaciers covered Europe and much of North America. The last period of glaciation ended about 18,000 years ago. After an initial rapid ice melt during which sea level rose rapidly for a few thousand years, the sea level has been rising very slowly at a constant rate for the last 10,000 years. Contrary to false media reports, sea level isn't increasing any faster now than it was a hundred years ago. The ocean level has risen only about seven inches in the past century, according to the National Oceanic and Atmospheric Administration (NOAA).[48]

Ice core data from Greenland's glaciers indicates that during most of the past 10,000 years Earth's temperatures were warmer than today. In the past 500 years there have been about twenty periods of warming and cooling.

Many climatologists believe that we're living in an interglacial period - between two parts of an ice age - the next part of which could resume within as short a time as a thousand years. During recorded history there was a warm period of time lasting from 250 BC to about 400 AD. This was called the Roman Warm Period. Modern coastlines have been found two miles away from where they were during the Roman invasion of Britain in 43 AD, strongly implying that at that time sea levels were significantly higher than they are now. Arctic temperatures were at least 2° higher then than now.[49] During that time the western Arctic ocean was covered in sea ice only about 50% of the time, whereas now it is ice covered for over ten months each year.[50]

---

47 Evans, op. cit.

48 https://www.ngs.noaa.gov/GRD/GPS/Projects/CB/SEALEVEL/sealevel.html

49 http://notrickszone.com/2018/04/26/arctic-temps-2-6c-warmer-than-today-with-4-5-fewer-months-of-sea-ice-coverage-2000-years-ago/#sthash.oUxXZuIP.dpbs

50 J.L. McKay, et al , Holocene fluctuations in Arctic sea-ice cover: dinocyst-based reconstructions for the eastern Chukchi Sea1-2

During the ensuing period from 400 AD to 950 AD, known as the Dark Ages, it was colder.

Then came the Medieval Warm Period, extending from about 950 AD to 1250 AD. Ice core samples show that although the average temperature was about two degrees warmer during those years, the carbon dioxide level was lower. This again demonstrates the lack of correlation between the percentage of carbon dioxide and Earth's average temperature. During the Medieval Warm Period grapes were grown in Britain. At the present time Britain is too cold to raise grapes. During that period in history a colony of Vikings from Scandinavia raised vegetables in a part of Greenland that is now covered by a glacier.

Empirical data contradicts the global alarmists' claim that the Medieval Warm Period wasn't a worldwide phenomenon. For one thing, a thousand years ago the snow line in the Rocky Mountains was 1000 feet higher than it now is, and forests in what is now Canada were growing north of where the tree line is now.[51] A 5000-year-old white spruce was found 100 km north of the current tree line in Canada. Permafrost now prevents trees from growing north of that line.[52] 1000-year-old tree stumps have also been found in ice caves under the Mendenhall glacier near Juneau, Alaska. There was a forest growing there 1000 years ago when it was warmer.

Then, around 1250 AD the temperature began to drop again, and by 1350 AD the Greenland settlement had been abandoned. Shortened growing seasons and torrential rains in Northern Europe led to the "Great Famine" of 1315-1317 AD.[53] Temperatures continued to decline, and a very cold period known as the Little Ice Age lasted until the late 1700's. Ever since that cold spell ended, the earth has been in a gradual warming trend. Average temperatures have increased about 0.5°C. per century ever since, and scientists really can't explain why.

---

51 Robert Zubrin, *Merchants of Despair*, New York: Encounter Books, 2012, p. 227

52 Ball, op. cit.

53 Larry Bell, "The New York Times' Global Warming Hysteria Ignores 17 Years Of Flat Global Temperatures," *Forbes Magazine*, Aug. 21,2013

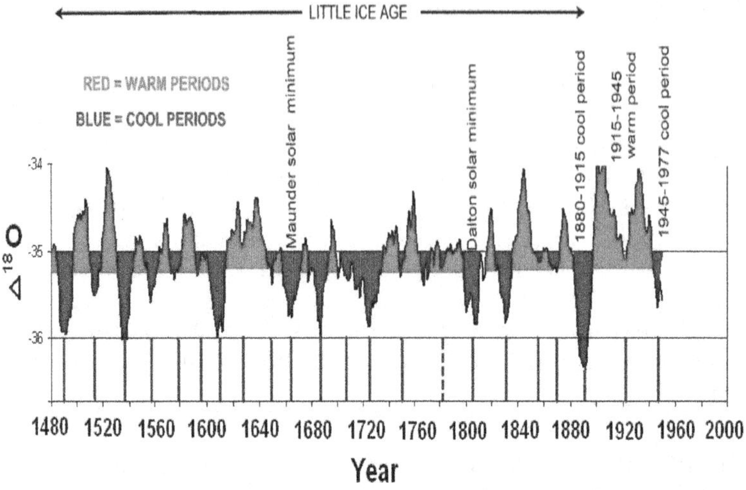

From the beginning of the 20th century until the present time there have been four periods of warming and cooling.

1. Between 1910 and 1940 average temperatures increased. At that time the level of CO_2 in the atmosphere was only 0.032% (320 ppm).

2. From 1940 to 1978 Earth was cooling at the same time that CO_2 levels were rising.

3. From 1978 to 1998 another period of warming occurred. CO_2 was also increasing during that time period, causing some scientists to mistakenly conclude that this was causing that temperature increase. But they evidently failed to note that the sun was unusually active during that 20-year period.

4. Due to a super strong El Niño, 1998 was unusually warm, causing a temperature spike. In El Niño conditions, a band of warm ocean water develops in the central and eastern part of the Pacific ocean. El Niños, lasting for a year or so, cause Earth to warm slightly. This weather phenomenon occurs about once every 8-10 years.

Between 1998 and 2019 average temperatures remained

fairly constant, perhaps cooling slightly, even though during this time period about 30% of all human generated CO_2 was emitted. In 2016 there was another El Niño, slightly raising the average temperature for that year.

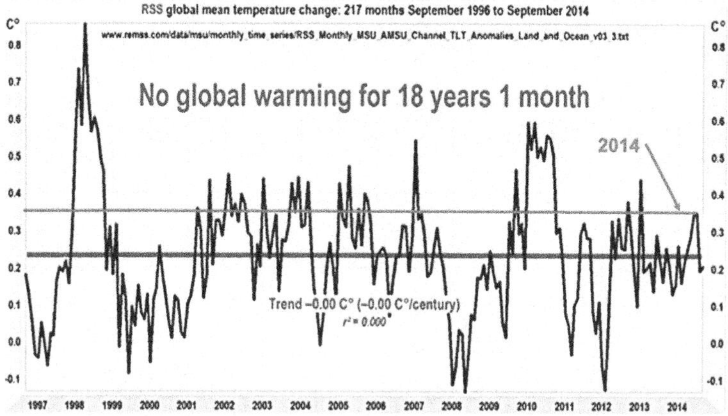

Of the above four time periods, #2 and #4 counter the human-caused global warming alarmist hypothesis. The decade of the 1930's, the warmest in the last century, occurred during period #1 when the carbon dioxide level was much lower than it now is. From 1940 to 1978 the earth was cooling while the CO_2 level was rising. From 1998 to the present the temperatures were fairly constant, while at the same time the CO_2 level was continuing to rise. 2013 was the coldest winter in 100 years in Europe. Five of the six snowiest winters occurred within the last decade.

For the first fifteen years of this century the IPCC, CRU and other agencies acknowledged that there was no global warming since 1998. Nevertheless, to everyone's surprise, in 2015 a paper by Tom Karl of NOAA, apparently timed to coincide with the upcoming Paris conference, claimed that the temperature data showing this lack of warming never existed and that temperatures had been rising during that time.

Dr. John Bates, a top NOAA climate scientist with an outstanding reputation,[54] came forward to reveal that this paper, which became known as the "Pausebuster Paper", was based on unverified and misleading data. Nevertheless, Karl's

54 He has since retired.

paper had a tremendous influence on policy makers from several nations who were trying to decide just how far to commit their governments in dealing with what they were told was a crisis.[55]

Temperatures in the Arctic were higher during the 1930's than they are now. Arctic ice rebounded during the cooling period from the late 1940's through the 1970's.

The period from October 2018 to April 2019 was the coldest and wettest 6-month period in the United States in over a century.[56]

55 Anthony Watts, "Bombshell - NOAA Whistleblower says Karl et al. "Pausebuster" Paper Was Hyped, Broke Procedures," WattsUpWithThat.com, Feb. 4, 2017
56 Tony Heller, YouTube video, "Coldest October-April In The US In Over A Century," May 2, 2019

5. Myths Perpetrated by Global Warming Alarmists

MYTH #1 Because of human-caused climate change the polar bear is now in danger of extinction.

THE FACTS The headline of an undated article on the World Wildlife Fund website reads, "Polar Bear Population Decline a Wakeup Call for Climate Change Action."[57] In reading this article we learn that it is the polar bear population in the Beaufort Sea - ***not*** the world polar bear population - that has declined.

In contrast, a 2011 article in International Business Times stated,

> "The US Fish and Wildlife Service estimates that the polar bear population is currently at 20,000 to 25,000 bears, up from as low as 5,000-10,000 bears in the 1950's and 1960's. A 2002 US Geological Survey of wildlife in the Arctic Refuge Coastal Plain noted that the polar bear populations 'may now be near historic highs'".[58]

This figure is based on research by J. Scott Armstrong of the University of Pennsylvania and others in a paper published in 2008. This number is verified by the IUCN/SSC Polar Bear Specialist Group.[59] Because of strict hunting regulations, the number of polar bears has been increasing for many years. "The Inuit are the only people in Canada who can legally harvest a Polar Bear based on cultural and subsistence purposes and is done through strictly enforced regulations and highly monitored tag allocations."[60] So we can put our fears to rest that polar bears are declining in number. That assertion just

57 https://www.worldwildlife.org/stories/polar-bear-population-decline-a-wake-up-call-for-climate-change-action

58 Amrutha Gayathri, "Polar Bear Population Higher than in 20th Century: Is Something Fishy about Extinction Fears? International Business Times, August 2, 2011.

59 IUCN/SSC Polar Bear Specialist Group, http://pbsg.npolar.no/en/index.html

60 http://www.canadanorthoutfitting.com/arctic-expeditions/polar-bear/

isn't correct.

It is no accident that the polar bear was chosen as the poster child of the global-warming-alarmist movement. From a distance they look cute and cuddly, but in actuality they are huge animals that have been known to attack humans. Their use as a symbol provides evidence of a cleverly devised propaganda campaign.

The Inuit people fear that there are now so many polar bears that they've jeopardized public safety. This is the conclusion of a four-year study. "Public safety concerns, combined with the effects of polar bears on other species, suggest that in many Nunavut communities, the polar bear may have exceeded the co-existence threshold."[61]

* * * * *

MYTH #2 The Antarctic ice shelf, Greenland glacier and Arctic sea ice are all melting.

THE FACTS: To depict an impending catastrophe, the first scene of a 2004 science fiction movie, "The Day After Tomorrow", showed a coastal portion of the West Antarctic ice sheet crashing into the ocean. But recently geologists at Washington University in St. Louis accidentally discovered the cause of that warming: a chain of active volcanoes under this glacier, after detecting unexplained seismic activity in that region.[62]

The uncontested fact, according to NASA, is that the total amount of ice in Antarctica has been growing for several decades.[63] This glacier, which contains about 90% of the earth's fresh water, is exceptionally stable. It is almost three miles thick (15,000 feet) in the center.

The average temperature in Antarctica is -58° F. To get ice to melt, you'd have to raise that temperature about 90° to above

61 Bob Weber, "Nunavut's polar bear population is unsafe, government document says," *The Canadian Press,* Nov. 12, 2018

62 https://www.investors.com/politics/editorials/active-volcanoes-melting-west-antarctic-ice/

63 "NASA - The Mystery of Antarctic Cooling," YouTube video, Jan. 25, 2016

the melting point of 32°F. Therefore, there have been no gaps in the Antarctic ice core record for the last 15 million years.[64]

The snow and ice at the South Pole has been accumulating so rapidly that the first two scientific observation stations had to be abandoned after being taken over by ice. The third station was constructed on stilts so that it could be raised as needed.[65] According to NASA, "The Antarctic ice sheet showed a net gain of 112 billion tons of ice a year from 1992 to 2001."[66]

Regarding Greenland: a plane that crash landed on the ice there in 1942 was recovered fifty years later in 1992 under 268 feet of ice. This demonstrates that Greenland's ice has been getting thicker, not thinner. The Greenland glacier is melting at the edges but getting thicker in the interior. Of course, every summer there is some melting of icebergs near the ocean. The following winter the melted ice is replaced. This has been occurring for thousands of years.

A large Greenland glacier, named Jakobshavn, which had been thinning out around 2012, started growing again in 2017, according to NASA's Oceans Melting Greenland (OMG) project. The study's lead author, Ala Khazendar, a NASA glaciologist on the OMG project, believes that cooling of the North Atlantic due to the shift in the North Atlantic Oscillation probably caused the glacier to rebound.[67] Other NASA scientists have called this a "miracle".

Because the North Pole is covered by ocean, there is no glacier there. It's covered by sea ice averaging about three meters in thickness. There have been warmer times in recorded history when ships were able to navigate through the Northwest Passage. As has been pointed out, temperatures in the Arctic were higher in the 1930's than today. For generations alarmists have been predicting that Arctic ice would soon disappear. All we have to do is look at the newspaper and magazine articles from the early 1900's that made that prediction.[68] As recently as

64 Professor Emeritus Don Easterbrook's eye-opening talk to the Washington State Senate committee on Climate Change, March 26, 2013

65 Ibid.

66 https://www.nasa.gov/feature/goddard/nasa-study-mass-gains-of-antarctic-ice-sheet-greater-than-losses

67 Chaffin Mitchell, "NASA study reveals astonishing growth of one Earth's previously fastest-shrinking glaciers," AccuWeather.com

68 Tony Heller, "The 'Ice-Free Arctic' Big Lie", https://realclimatescience.com/

2008 scientists were predicting that the Arctic would be ice free by 2013. Not only did this not happen but now it is increasing

A good example of a half-truth so common in our media is the following statement: "Melting in the Arctic reached an all-time high in June. Ice has been disappearing at 29,000 square miles a day."[69] Dr. Tim Ball explained what is really happening here: "This is near the average daily rate of melt in the brief Arctic summer, but few people know this is natural. Approximately ten million km^2 melts every summer in approximately 145 days."[70] Usually all or more of this melted ice is replaced the following winter.

From 1880 to 1915 glaciers advanced. Then they receded again during the period from 1915 to 1945. They started advancing again from 1945 to 1977 when Earth was again cooling.

* * * * *

MYTH #3 A 97% consensus [agreement] of scientists say that Earth's climate is experiencing dangerous warming caused by human activities.

THE FACTS This erroneous and totally misleading figure was obtained from four different studies, all of which were flawed.

The first was done in 2004 by Naomi Oreskes, who is not a scientist. She examined the abstracts [summaries] of 928 scientific studies that she had found by putting in the search term "global climate change". This yielded only 8% of the results she would have obtained from the search term "climate change", thus leaving out hundreds of studies focusing on natural causes of climate change. The topics of many articles counted in this study didn't address causes of climate change but, assuming that IPCC conclusions were correct, focused on the influence of climate on the incidence of influenza, the life cycle of frogs, etc. Before publishing Ms Oreskes' article the journal editor failed to check either her methodology or her

---

2016/10/the-ice-free-arctic-big-lie/, Oct. 14, 2016

69 *The Daily Mail*, July 2016

70 Dr. Tim Ball, *Human Caused Global Warming: The Biggest Deception in History*, 2016, Tellwell Talent, p. 8

sources. Shortly thereafter Naomi Oreskes wrote a book entitled, *Merchants of Doubt,* which lambasted climate skeptics.

In a 2009 study (Doran and Zimmerman), Maggie Zimmerman, a graduate student, sent a 2-minute online survey to 10,257 employees of schools and government research agencies. This left out over 10,000 geologists, physicists, meteorologists and astronomers, many of whom were studying the issue. She received 3,146 responses. Yet her 98% figure was based on only 79 responses she had hand picked - certainly not a representative sample. So this study has been debunked.[71]

In 2010 another college student, William Anderegg, identified 908 scientists who had written the most papers about global warming. He considered these to be most qualified to hold an opinion on climate change. "The 50 most prolific alarmists were published an average of 408 times each, versus only 89 times each for the skeptics."[72] Hundreds of the studies considered had several - as many as twelve - different authors.[73] This misinformation weighted the results against the skeptics. Anderegg failed to consider that the quantity of published studies doesn't equal correctness. Studies which reach conclusions agreeing with the official government position on climate change are heavily financed. In contrast, those scientists who study the natural causes of global warming must scramble for financing. This is one reason why these scientists aren't publishing as many peer-reviewed papers.

In 2013 John Cook, who is not a scientist but a professional cartoonist, issued a report to the Global Warming Policy Foundation stating that 97.1% of the 11,944 scientific papers he examined "explicitly or implicitly suggested that human activity was responsible for some warming."[74] This claim has been debunked by at least four other studies,[75] which found that this definition doesn't support the IPCC claim that mankind is responsible for most global warming. A study by Legates found that only 41 of these 11,944 papers stated the opinion that most

71 Craig Idso, et al, *Why Scientists Disagree About Global Warming,"* The Heartland Institute, 2016, pp.13-14

72 Ibid, p. 16

73 R.L. Holz, "How Many Scientists Does it Take to Write a Paper? Apparently Thousands," *The Wall Street Journal,* August 10, 2015

74 Idso, op. cit., p. 18

75 Tol (2013), Montford (2013), Friends of Science (2014) and Legates et al (2015)

of the warming since 1950 was manmade. It was revealed that Cook had told the publisher of his study its results (97% consensus) *before he had even done the research*.[76]

But science doesn't operate by consensus. To adhere to the scientific method, scientists are obligated to question all new ideas. In that case, even if this 97% figure were correct (it isn't), it would be meaningless. When Galileo challenged the scientific consensus that the earth was the center of the universe, he was met with an extremely hostile reaction.

Likewise, Dr. Barry Marshall, who hypothesized that ulcers are caused by the bacterium Helicobacter pylori, was challenged by the majority, who believed ulcers are caused only by stress. Finally Dr. Marshall empirically proved his hypothesis by drinking a solution containing these bacteria. In doing so he successfully gave himself an ulcer. For this he was awarded a Nobel prize.

In the early 20th century Alfred Wegener was scorned and mocked by other geologists for advancing the theory of continental drift, which led to the tectonic plate theory. This theory, the acceptance of which he didn't live to see, revolutionized the science of geology.

Many scientists who had once accepted the global warming hypothesis changed their minds after new information became evident. Klaus Eckert Puls, a German meteorologist, lamented:

> "Ten years ago I simply parroted what the IPCC told us. One day I started checking the facts and data. At first I started with a sense of doubt, but then I became outraged when I discovered that what the IPCC and the media were telling us was sheer nonsense and was not even supported by any scientific facts and measurements. To this day I feel shame that as a scientist I made presentations of their science without first checking it. Scientifically it is sheer absurdity to think we can get a nice climate by turning a CO_2 adjustment knob."[77]

76 Idso, op. cit.
77 Dr. Tim Ball, *The Deliberate Corruption of Climate Science, http://www.stairwaypress.com/bookstore/the-deliberate-corruption-of-climate-science/*

In the 2007 Global Warming Petition project 31,487 scientists, over 9000 of whom have PhD's, signed the following declaration:

> "There is no convincing scientific evidence that human release of carbon dioxide, methane or other greenhouse gases is causing, or will, in the foreseeable future, cause catastrophic heating of the Earth's atmosphere and disruption of the Earth's climate. Moreover, there is substantial scientific evidence that increases in atmospheric carbon dioxide produce many beneficial effects upon the natural plant and animal environments of the earth."[78]

For speaking out against the fraudulent AGW hypothesis a Canadian scientist, Dr. Tim Ball, was sued by three IPCC scientists who all had the same lawyer. In this way, the climate alarmist cabal had hoped to silence other scientists daring to speak out against their hypothesis through fear of the same consequence. Defending oneself against such lawsuits, no matter how frivolous or unjustified, often takes years and hundreds of thousands of dollars that could quickly use up one's life savings. Fortunately Dr. Ball has been able to fend off these unjustified lawsuits.

Scientists choosing to investigate natural rather than manmade causes of warming quickly learn that they will not be awarded government research grants. This is because government officials, lusting for carbon taxes, are interested only in those reports that can show in some way that humans are causing this so-called "problem". For example, if scientists can demonstrate that a species of butterfly has migrated north or is in danger of extinction, attributing this to human-caused climate change will guarantee them lucrative grants that are perpetually renewed.

Many scientists have been afraid to disagree with those in power out of fear that their careers will be ruined. Since the global warming hypothesis surfaced about 40 years ago, many meteorologists and climatologists have waited until after they retired to speak up. Others became more skeptical after more information surfaced. Just a few of the top scientists who are

78 http://www.petitionproject.org

now standing up and saying this is wrong are: Dr. David Evans, Dr. Joanna Simpson, Dr. Judith Curry, Dr. Alrude Allegre, Dr. Tafd Murty, Dr. Chris de Freitas, Dr. Kiminori Itoh, and Dr. Andrei Kapitsa.[79] You need to Google the names of these scientists to learn about why they've changed their stance.

Piers Corbyn, a well known and respected British meteorologist with the best record of accurate long range weather forecasting (WeatherAction.com) points out that the world is now cooling, not warming. He contends that

> "There is no observational evidence in the thousands and millions of years of data that changes in CO_2 have any observable effect on weather or climate in the real world. There are no scientists in the world who can produce such observational data, and we challenge anyone reading this to send us such observational evidence from the real world or find someone who can and get them to produce it. There is only effect the other way, namely that ocean temperatures control average CO_2 levels due to basic laws of physics about the solubility of CO_2 gas in (sea) water."[80]

Corbyn believes that because the low incidence of sunspots we are seeing now is similar to the Maunder Minimum from 1645-1715 and the Dalton Minimum from 1790-1830, we could well be at the beginning of another mini ice age, a prolonged period of cooling which could last for thirty years or more. This would be bad news for the world. It would reduce agricultural production, raise food prices and negatively impact the economy in other ways.

In 2012 forty nine former NASA scientists, including seven astronauts, sent a letter to NASA expressing their dismay that this agency - specifically the Goddard Institute for Space Studies (GISS) - was drawing conclusions based on scientifically flawed climate models that had failed to make accurate predictions and was ignoring more scientific methods of investigation.[81]

79 U.S. Senate Minority Report
80 Piers Corbyn, www.WeatherAction.com.
81 Cfact, "Astronauts and scientists send letter to NASA: Stop global warming advocacy," www.cfact.org, April 10, 2012

Another prominent scientist, Hal Lewis, resigned from the American Physical Society (APS) in 2010 after noting that this society had been taken over by the unscientific global warming hysteria. In his letter of resignation, he stated:

> "The global warming scam, with the (literally) trillions of dollars driving it, that has corrupted so many scientists has carried APS before it like a rogue wave. It is the greatest and most successful fraud I have seen in my long life as a physicist."[82]

* * * * *

MYTH #4 Due to human caused global warming, the sea level is rising rapidly at an accelerated rate, and as a result many coastal areas will soon be under water. We are constantly told the following: "Global sea level is on the rise at an increasing rate, according to a new study. By the end of the century, it could rise another 2 feet."[83]

THE FACTS This map appeared in an issue of *National Geographic* magazine. The projected sea level rise used in this map is based on a sea level increase of 262 feet that was predicted in 2000 by the US Geological Survey (USGS). But such a sea level increase could only occur...

82 Hal Lewis, "Hal Lewis Resignation Letter From the American Physical Society," https:// www.heartland.org, Dec. 8, 2010

83 Brandon Miller, "Satellite observations show sea levels rising, and climate change is accelerating it," https://www.cnn.com/2018/02/12/world/sea-level-rise-accelerating/index.html, March 13, 2018

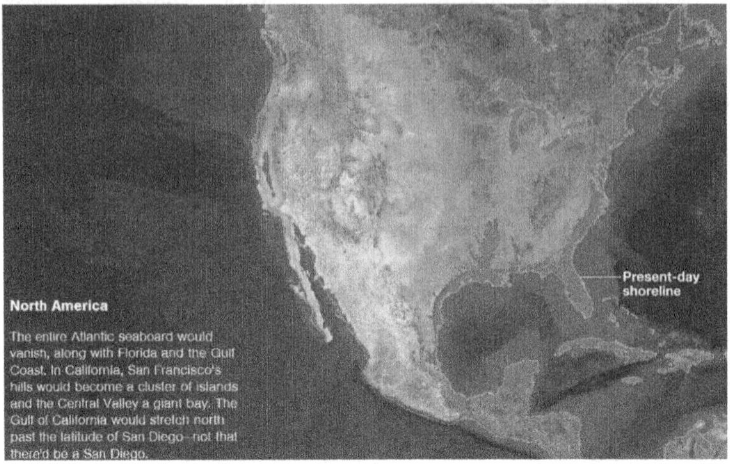

- If ***all*** of Alaska's glaciers melted. In that case sea level would rise ~0.05 meters (about 0.16 feet).

- If ***all*** of Earth's temperate glaciers melted. That would cause the sea level to rise ~0.3 meters (about one foot).

- If ***all*** of Greenland's glaciers melted, sea level would rise ~6 meters (or 19.7 feet).

- If ***all*** of Antarctica's glaciers melted, the earth's sea level would rise ~73 meters (about 240 feet).[84]

In other words, such a phenomenon would occur only if ***all*** of the ice on Earth were to melt. Although there have been periods in Earth's history when there were no glaciers or ice caps, nothing like that has occurred for at least fifteen million years. The periods of warming that have occurred in the last few hundred years, within natural limits, have provided no evidence that this could occur for a long, long time. So this picture in *National Geographic* magazine is a figment of the imagination and obviously another scare tactic. The editors of *National Geographic* should be ashamed that they ever published such a picture.

The fact is that the earth's sea level has been slowly rising at a

84 Philip Bump, "This Map of Sea Level Rise Is Probably Wrong. It's Too Optimistic," https://www.theatlantic.com/politics/archive/2013/11/map-sea-level-rise-probably-wrong-its-too-optimistic/354766/, Nov. 4, 2013

constant rate for the last few thousand years. Recently Nils Axel Mörner, a UN IPCC scientist who is a world-renowned expert on global sea level measurement, warned the IPCC that it was publishing false information which would inevitably be found to be incorrect. When they ignored him rather than heeding this warning, he resigned. Dr. Mörner has affirmed that Earth's climate is cooling and that ocean levels will decrease.[85]

He states that the sea level has been rising at an average rate of only 1.1 mm - about the thickness of your fingernail - per year. This figure is confirmed by NOAA.[86] It amounts to about seven inches per century, not the 1-2 meters per century predicted by some government agencies.

Sea level is tricky to measure, because the level of the land can be rising or sinking, a phenomenon called isostasy. For example, freed from the weight of glaciers that covered it in the last ice age, Scotland is rising. Meanwhile, Miami, Florida, is sinking partly because of the removal of so much water from the water table and the construction of many large buildings close to the shoreline.

Currently the northern end of the North American continent is rising, due to the melting of the mile-thick Laurentide glacier that covered it 20,000 years ago. When that glacier melted between 15,000 and 8,000 years ago, the land it had covered slowly started rising, causing the southern end of the continent, bordering the Gulf of Mexico - including the Miami area - to sink.[87]

These facts are simply ignored by the alarmists. So we still see statements in articles such as the following: "The rate of sea-level rise has tripled over the last decade, according to a recent study from the University of Miami, bringing with it more frequent coastal flooding."[88] But this is because Miami and the surrounding area are sinking due to isostasy and other causes.

---

85 Alex Newman, "UN IPCC Scientist Blows Whistle on Lies About Climate, Sea Level," *New American*, Feb. 12, 2019

86 https://www.noaa.gov/explainers/tracking-sea-level-rise-and-fall

87 Dr. Tim Ball, "Sea Level Rise; A Major Non-Existent Threat Exploited by Alarmists and Politicians", WattsUpWithThat.com, July 21, 2018

88 Chris Bentley, "Facing rising sea levels and greater insurance risk, Southern Florida braces for relocations, new flood design standards, and more," https://archpaper.com/2017/04/south-florida-rising-sea-levels, April 12, 2017

Sea level measurements are made both by averaging tidal gauge data and by satellite altimetry. It was reported to Dr. Mörner that the satellite data had been tampered with to show a rise in sea level.[89]

Another system called GRACE – made up of gravitational-anomaly satellites - computes the mass of the ocean, from which changes in sea-level can be directly calculated. According to these data sea level fell slightly from 2002-2007.

Media reports falsely claim that, due to the rising sea level, the Maldives Islands in the Indian Ocean are being submerged underwater and that its people will thus have to be relocated. However, Dr. Mörner stated,

> "Clear observational measurements in the field indicate that sea level is not rising in the Maldives, Bangladesh, Tuvalu, Vanuatu, and French Guyana. (Mörner, 2007abc, 2010ab). All these are key sites in the sea level debate, where the IPCC and its ideological associates have predicted terrible flooding."[90]

A group of Australian environmental scientists uprooted a 50-year-old tree that was near the shore, because it was a landmark showing where the sea level had been. There are too many such instances of dishonesty to promote a political – not scientific - agenda.[91]

> "The president of Tuvalu continues to claim that his islands are being flooded. Yet the tidal-gauge data provide clear indications of stability over the last 30 years.[92] In Vanuatu, the tidal gauge also indicates a stable sea level over the last 14 years."[93]

---

89 Nils Axel Mörner, http://scienceandpublicpolicy.org/images/stories/papers/reprint/sea_level_ not_rising.pdf
90 Ibid.
91 Ibid.
92 Mörner, 2007ac, 2010b; Murphy, 2007
93 Ibid.

These graphs show that the rate of sea level increase has not

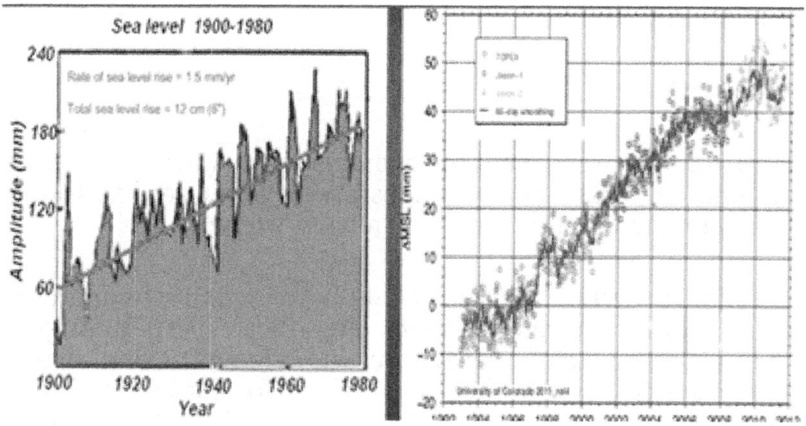

increased but has actually slowed slightly in the past decade.

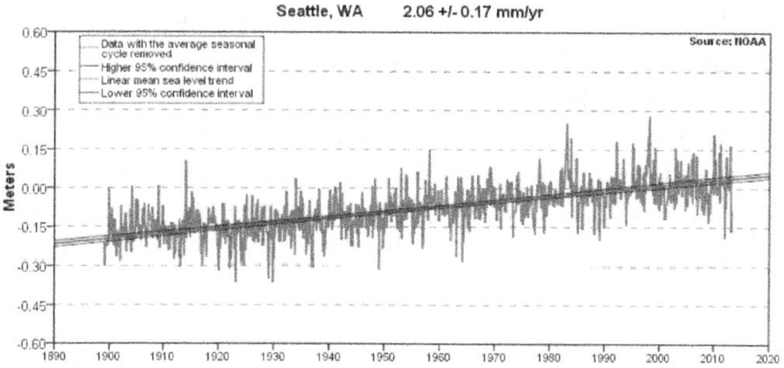

Dr. Mörner points out that the sun is the main driver of climate. At the present time the sun has very few sunspots, and, as a result, Mars, Jupiter and Saturn are also showing the effects of decreased solar activity.

He says there's something "sick" in the IPCC's methods. "In science we discuss. We don't forbid or neglect...They are working with lobbyists. They blacklist those who disagree."[94] As with everything they do, the IPCC starts with the assumption that their estimates of global warming are accurate and therefore their extrapolation of perceived trend going forward of increased sea level is also accurate. Attributing 50 percent of the increased sea level to expansion of the ocean due to what is

94 Newman, op. cit.

done with a computer model that eliminates almost all the variables that cause temperature change is extremely arrogant and foolhardy.

* * * * *

MYTH #5 Extreme weather events such as hurricanes, floods, droughts, wildfires and tornadoes are a direct result of human caused climate change, which is getting worse.

THE FACTS When, following 1998, the Earth's temperatures ceased increasing and leveled out, those perpetuating the global warming scare seized on the idea of using extreme weather events, which have been occurring for all of recorded history, to claim that they've been getting worse or are happening more often. The public is continuously bombarded with media stories of such events.

Dr. Roger Pielke, Jr., whose degrees are in mathematics, public policy, and political science, began studying extreme weather and climate in 1991 at the National Center for Atmospheric Research in Boulder, Colorado. Although he isn't a climate change skeptic, Dr. Pielke has provided irrefutable statistical evidence clearly demonstrating that these claims are false. In testimony before the U.S. Senate Committee on Environment & Public Works, he laid out the documented facts, summarized as follows.

> "It is misleading, and just plain incorrect, to claim that disasters associated with hurricanes, tornadoes, floods or droughts have increased on climate timescales either in the United States or globally. It is further incorrect to associate the increasing costs of disasters with the emission of greenhouse gases.
>
> "Globally, weather-related losses ($) have not increased since 1990 as a proportion of GDP (they have actually decreased by about 25%) and insured catastrophe losses have not increased as a proportion of GDP since 1960.
>
> "Hurricanes have not increased in the US in frequency, intensity or normalized damage since at

least 1900. The same holds for tropical cyclones globally since at least 1970 (when data allows for a global perspective).

"Floods have not increased in the US in frequency or intensity since at least 1950. Flood losses as a percentage of US GDP have dropped by about 75% since 1940.

"Tornadoes have not increased in frequency, intensity or normalized damage since 1950, and there is some evidence to suggest that they have actually declined.

"Drought has for the most part, become shorter, less frequent, and covers a smaller portion of the US over the last century." Globally, there has been little change in drought over the past 60 years."[95]

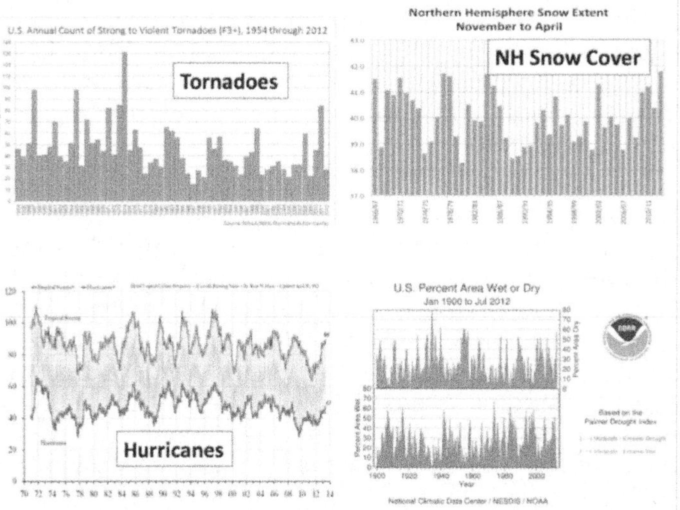

These carefully compiled statistics were part of Dr. Pielke's sworn testimony before Congress. Because they couldn't be disputed, global warming alarmists pulled out all stops to viciously attack his credentials, character and reputation. They claimed that he was "outside of the science community" and "on the fringe." The University of Colorado, where he teaches, came to his defense.

---

95 Dr. Roger Pielke, Jr., Congressional testimony, July 18, 2013

AGW alarmists frequently cite Hurricane Katrina and superstorm Sandy, as "proof of the effects of climate change." Katrina, which made landfall on Aug. 29, 2005, was a truly wicked category-5 storm, but it was far from the worst hurricane ever experienced. What caused so much damage was that the levees built in New Orleans to prevent storm surges from moving onto the land, broke. Most of that damage was caused by the resulting flooding. These levees hadn't been properly maintained.

> "The catastrophic failure of New Orleans's hurricane protection system represents one of the nation's worst disasters ever...A large proportion of the destruction from Hurricane Katrina was caused not only by the storm itself, however, but also by the storm's exposure to engineering and engineering-related policy failures."[96]

Sandy was just a category-1 hurricane. But it did a great deal of damage due to a bizarre set of circumstances. First, as it moved up the east coast of the US in late October of 2012, it slammed into an extremely cold air mass - called a noreaster - moving south from Canada. This cold air mass added a great deal of energy to the system.

Sandy made landfall just south of Atlantic City, New Jersey, at the height of a spring tide, an unusually high tide that occurs when the sun and moon are lined up in such a way that they are both pulling in the same direction. "The storm surge - an additional 9 feet or more of water piled up against the coast by furious winds and crashing ocean waves - was riding on top of this extra high tide."[97] Areas that were hardly impacted at all by this storm experienced mysterious power failures. Many weather stations in the wider area also mysteriously went offline.

The ten worst hurricanes in US history, listed in order of their severity and the damage and loss of life they caused are:

---

96 https://eng-resources.uncc.edu/failurecasestudies/dam-failure-cases/new-orleans-hurricane-katrina-levee-failures/ (ASCE Review Panel 2007, p. v)

97 Malcolm J. Bowman, "Superstorm Sandy - How did it happen and are we prepared for the future?"https://seagrant.sunysb.edu/media/sandy12/UUPInsight-Sandy020113.pdf

1900 Great Galveston Hurricane

1928 OkeeChobee Hurricane

2005 Hurricane Katrina

1893 Chenière Caminada Hurricane

1893 Sea Islands Hurricane

1881 Georgia/South Carolina Hurricane

1967 Hurricane Audrey

1935 Great Labor Day Hurricane

1856 Last Island Hurricane

1926 Great Miami Hurricane

As you can see, the great majority of these hurricanes occurred before the level of carbon dioxide in the atmosphere began to rise.[98] In fact, of these ten the only recent ones were Hurricane Audrey in 1967 and Hurricane Katrina in 2005. So the continual torrent of climate change alarmist reports issued every time there is a hurricane, tropical storm or flood is nothing but a farce.

* * * * *

MYTH #6 2004 was the beginning of an intense propaganda campaign which stated, in effect, "The oceans are becoming more acidic as a result of all the carbon dioxide man has put into the atmosphere. This will lead to the extinction of all marine species starting with photoplankton at the bottom of the marine food chain."

THE FACTS Just the opposite is true. Due to the warming

98 Kim Bryan, "10 Worst Hurricanes in United States History," https://owlcation.com/stem/Worst-Hurricanes-United-States-History, Jan. 6, 2017

trend that has been occurring since the Little Ice Age ended in the late 1700's, the oceans have been gradually warming. Since more carbon dioxide can dissolve in cold water than in warm water, some of the vast quantity of carbon dioxide dissolved in the oceans has come out of solution into the atmosphere.

Here's an experiment you can do that demonstrates this scientific fact:

Pour out two glasses of room temperature soda pop - which is CO_2 dissolved in water, syrup and flavorings – one of which contains ice cubes and the other which doesn't. In fifteen minutes compare the pop in both glasses. The one with the ice cubes will still have some fizz. The other will be flat. More CO_2 remains in the glass of colder soda pop. This shows that colder water is able to hold more carbon dioxide than warmer water.

When ocean water warms, it can't hold as much carbon dioxide, so some of it is released. This is a natural cause of the increased CO_2 in the atmosphere. So, rather than there being more carbon dioxide in warmer ocean water, there is less.

Carbon dioxide reacts with water to form a weak acid [H_2CO_3: carbonic acid]. So when CO_2 comes out of solution, there's less, not more, carbonic acid in the water. Therefore, as the ocean warms, releasing CO_2, it becomes less, not more, acidic. Neutral liquids, containing equal amounts of H+ and OH- ions, are designated a pH of 7.0 on the pH (potential hydrogen) scale. An acid is any liquid having a pH of less than 7.0. A base has a pH that's higher than 7.0. The pH of ocean water ranges from 7.5 to 8.3. Therefore, ocean water is basic, and it is extremely unlikely that it could become acidic for many thousands of years.

Due to the unequal heating of water at different latitudes, ocean currents are constantly circulating. Colder water sinks and warmer water rises, since colder water is more dense than warmer water. In an uncommon case where there is upwelling and cold water finds its way to the surface, or if there is a tidal pool into which acid from the land is flowing, the pH in that area could become slightly reduced. But in no case would ocean water in general ever become close to acidic.[99]

So this idea that ocean water is becoming acidic as a result of climate change is nothing but another falsehood perpetrated by fear mongers for the purpose of frightening the public.

* * * * *

MYTH #7 The extremely cold winter weather and snow recently experienced in the temperate zones are caused by global warming. This has created a polar vortex in the Arctic that drives frigid Arctic temperatures south and warmer temperatures into the Arctic regions.

THE FACTS We often see headlines such as "Polar Vortex to Freeze US Upper Midwest and Northeast".[100] In this article "polar vortex" is incorrectly defined as "the frigid winds that circulate around the North Pole."

The misleading term "polar vortex," coined in the 1800's, was seized upon by Eric Holder, the Obama administration's science czar. Its purpose is to convince people that cold weather in the temperate zones is somehow caused by manmade global warming. However, a vortex is actually a rapidly rising and spinning column of warm air in the center of a hurricane or tornado. There is no such thing as a "polar vortex".

What Holder was really referring to is a ***jet stream***, a high altitude, 30,000 ft, air current rapidly moving from west to east. Jet streams zigzag from north to south in a cyclical fashion.

Occasionally this jet stream loosens, causing a greater variation from north to south. This brings frigid Arctic air south to temperate latitudes, causing extremely cold conditions. According to meteorologist Piers Corbyn, such a phenomenon is more common when the earth is in a ***cooling*** - not warming -

---

99 Dr. Don Easterbrook, lecture before Washington state senate committee, March 26, 2013
100 CNBC, Jan. 29, 2019

phase.[101]

* * * * *

MYTH #8 "The impacts of climate change are already being felt in communities across the country."[102]

THE FACTS This statement raises a red flag demonstrating shameful propaganda, even from the government. We need only to look at the evidence to realize that this is a false claim. Despite what you may have heard on TV or read in articles:

> (1) Satellite temperature records indicate that there has been *no warming at all* in the past 20 years.
>
> (2) Statistics show that the incidence of serious hurricanes, wildfires, tornadoes, drought, and flooding is about the same or even less than before the percentage of atmospheric carbon dioxide increased.
>
> (3) The rate of sea level rise is about the same as it has been for the last 1000 years.
>
> (4) There is no physical evidence that the pH of the ocean has been reduced. The ocean is not acidic but alkaline.

Hurricane Katrina, an extremely destructive and deadly category-5 hurricane that made landfall in 2005 on Florida and Louisiana, is often used as a case in point. It was featured heavily in Al Gore's documentary in 2006, just a year later. This hurricane was still fresh in everyone's memory, giving the film a tremendous impact. However, looking at the entire record, we know that the incidence of major hurricanes has actually decreased in recent years.

---

101 Piers Corbyn, "Secrets of the Polar Vortex Exposed", video, Jan. 19, 2014.
102 Fourth National Assessment,"Volume II: Impacts, Risks, and Adaptation in the United States," https://nca2018.globalchange.gov

6. The Politics of Global Warming

A Canadian oilman who made a huge fortune in the US, Maurice Strong played the most important role in establishing the Intergovernmental Panel on Climate Change (IPCC) in 1988. A friend of the oil magnate David Rockefeller, Strong was appointed chairman of the 1972 Earth Day UN Conference, at which he advocated population reduction and the lowering of living standards in the interest of "saving the environment." His radical viewpoints were revealed years later when he asked, "Isn't the only hope for the planet that the industrialized civilizations collapse? Isn't it our responsibility to bring that about?"[103]

Along with the Rockefellers, Strong had helped found the Club of Rome in 1968. The members of this organization believed that the earth's resources were being used up too fast, especially by the industrialized nations. Their beliefs stemmed from the mistaken idea advanced by Thomas Malthus in his 1798 *Essay on the Principle of Population* that the world's population would outgrow the food supply. His conclusions were disproved by what actually happened when the world's population increased.

Malthus' deluded philosophy contributed to the ideas of Charles Darwin[104] and the development of the eugenics movement in the early 1900's. This movement, also based on warped thinking, advocated the selective breeding of the human population - by methods such as the sterilization of those deemed inferior - to improve its genetic composition. But who would decide who was superior and who was inferior? Obviously it would be those in power. This led to the unfortunate episode of the Holocaust by the Nazi's in World War II.

In his book *The Population Bomb,* published in 1968, Paul Ehrlich revived Malthus' idea. In this book Ehrlich predicted, "The battle to feed humanity is over. In the 1970's, the world

---

103 WikiQuote.com/wiki/Maurice_Strong

104 The complete name of his famous book is *The Origin of the Species by Means of Natural Selection, or the Preservation of Favoured Races in the Struggle for Life.*

will undergo famines. Hundreds of millions of people are going to starve to death in spite of any crash programs embarked upon now. Population control is the only answer."

Time has shown all of Ehrlich's forecasts to be incorrect. Yet there are still many people, some of whom are billionaires, who believe the world's population should be as low as one billion - not the current 7.5 billion. They're afraid that if third-world people get the energy they need to raise their living standard, the earth's population will rise further.

Actually the opposite is true. History has shown in every instance that when a population's living standards increase, the birthrate decreases.[105] This principle is known as the Demographic Transition. History has shown repeatedly that industrialized civilizations have lower birthrates and death rates, as the following chart demonstrates.

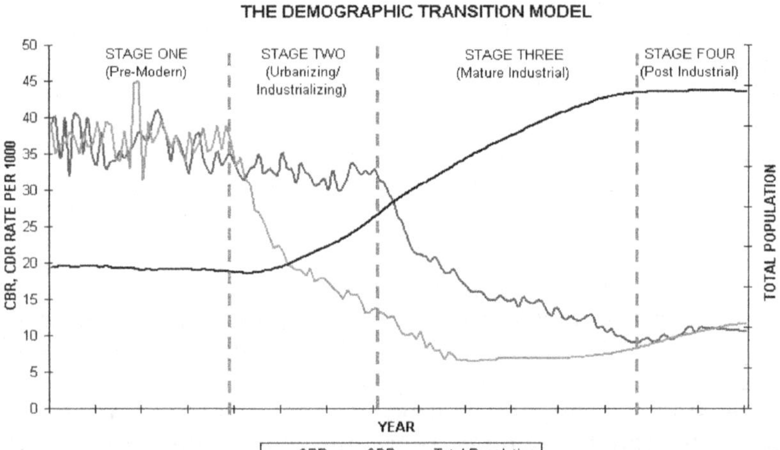

The Club of Rome's frightening goal and strategy based on these mistaken ideas was revealed in its book *The First Global Revolution (1991),* which asserts, "In searching for a new enemy to unite us, we came up with the idea that pollution, the threat of global warming, water shortages, famine and the like would fit the bill...All these dangers are caused by human intervention...The real enemy, then, is humanity itself."[106] Here

105 Ball, op. cit., p. 95

106 Alexander King and Bertrand Schneider, *The First Global Revolution,* 1991, Pantheon Books

they had hit on something for which to make people feel guilty.

A well known strategy for grabbing power that has been known and used for a long time is -

PROBLEM → REACTION → SOLUTION

In other words, those in power create a problem or crisis that is real or imagined. They sell this "problem" to the people using thousands of reports and articles - cleverly contrived propaganda. The public reacts to this problem by begging for a solution, which is then provided handily by the people in power. Because global warming is a worldwide "problem", the "solution" is too difficult for nation states by themselves to handle. Thus, the solution to this particular "problem" is nothing short of totalitarianism on a global scale.

Just as in the book, *1984,* a common enemy and a common cause around which the entire planet could rally had been found. About 80% of Americans identify themselves as being deeply concerned about the environment. It has become evident that this important cause has been taken over and used by those who have a great deal to gain from it. In reality the ruling elite have no interest in the environment, as demonstrated by their huge mansions and private jets in which they regularly fly to exotic places for climate conferences.

Politics is all about control. Taking the moral high ground, the power seekers have claimed that only those advocating action against human-caused global warming care about the earth. In so doing they've literally hijacked the environmental movement.

After the first Earth Summit, Strong was appointed Assistant Secretary General of the United Nations. In 1992 he organized the UN Framework Convention on Climate Change (UNFCCC) at the Rio Earth Summit. It was at this conference, of which he was chairman, that the globalist-inspired UN Agenda 21 (which has now morphed into "Agenda 2030") was also established.[107]

It's no accident that the UNFCCC and the Agenda 21 plan were

107 Engdahl, op. cit.

given birth at the same time. The UNFCCC and Intergovernmental Panel on Climate Change would provide the excuse for the establishment of Agenda 21.

When the investigative journalist, Elaine Dewar, author of *Cloak of Green (1995),* interviewed Maurice Strong, she asked him why he didn't run for office. Strong responded that he could get his agenda accomplished much more easily at the UN, where he could get all the money he needed, appoint anyone he wanted and push his ideas forward without interference. Her conclusion was, "Strong was using the UN as a platform to sell a global environment crisis and the global governance agenda."[108]

On the scientific front Strong set up the United Nations Environmental Program (UNEP) and the World Meteorological Organization (WMO) to create the science he needed to advance the Club of Rome's political agenda. The Intergovernmental Panel on Climate Change (IPCC) would provide policy recommendations for politicians. Climate scientists were appointed to this agency to research the subject and write reports, each of which would lend the scientific authority for the policy reports to be handed to individual governments.

Here's how Hubert Lamb (1913-1997), founder of the Climate Research Unit in East Anglia, in Great Britain, described him:

> "Strong prefers to operate in the background. He, perhaps more than any other single person, is responsible for the development of a global agenda now being implemented throughout the world...The fox has been given the assignment, and all the tools necessary, to repair the henhouse to his liking."[109]

The UNFCCC literally controlled the outcome of this research by narrowing its focus. In their research the IPCC scientists were constrained to consider ***only*** the IPCC's definition of the term "climate change", which had been defined as:

> "a change of climate which is attributed directly or

108 Ball, op. cit., p.102
109 Ibid

> indirectly to *human activity* that alters the composition of the global atmosphere and which is, in addition to natural climate variability, observed over considerable time periods."[110]

In writing their reports, IPCC scientists were forced to operate only in the confines of this definition and not to any alternative possibilities. The IPCC would write its summary report *before* the scientific report had even been submitted. They'd take that report to the scientists and tell them, "Make sure your report agrees with this."

In their 1995 report the IPCC scientists summarized their findings as follows:

> "None of the studies cited above has shown clear evidence that we can attribute the observed [climate] changes to the specific cause of increases in greenhouse gases. While some of the pattern-base discussed here have claimed detection of a significant climate change, no study to date has positively attributed all or part of climate change observed to man-made causes. Any claims of positive detection and attribution of significant climate change are likely to remain controversial until uncertainties in the total natural variability of the climate system are reduced. While none of these studies has specifically considered the attribution issue, they often draw some attribution conclusions, for which there is little justification."

These scientists were thus truthfully acknowledging that although the climate was changing, as it always had, there was no direct evidence that recent changes had been caused by mankind.

However, despite this conclusion by the IPCC scientists, the final policy report summary composed by Dr. Ben Santer reads:

> "There is evidence of an emerging pattern of climate response to forcing by greenhouse gases and sulfate aerosols...from the geographical, seasonal and

110 Ball, Ibid.

> vertical patterns of temperature change...These results point toward a human influence on global climate.
>
> "The body of statistical evidence in chapter 8, when examined in the context of our physical understanding of the climate system, now points to a discernable human influence on the global climate."[111]

In writing this, Dr. Santer was completely disregarding the conclusions reached by the IPCC scientists.

Although many scientists who work for the IPCC are competent and truthful, few have actually had any influence over final IPCC conclusions. Their work is cherry picked line by line by political appointees, many of whom are not scientists at all but activists from environmental groups such as the World Wildlife Fund. These Summaries for Policymakers always end up saying what the IPCC wanted them to say all along and are then passed on to the media and the public.

> "IPCC's 1996 report used selective data, a doctored graph, and featured changes in text that were made after the reviewing scientists approved it and before it was printed. The many irregularities provoked Dr. Frederick Seitz, a world-famous physicist and former president of the US National Academy of Sciences, the American Physical Society, and Rockefeller University, to write (in August 1996) in the Wall Street Journal: 'I have never witnessed a more disturbing corruption of the peer review process than events that led to this IPCC report.'"[112]

Other top scientists who had formerly supported the IPCC have protested its unscientific practices. Professor Fritz Vahrenholt, a German environmentalist, criticized the IPCC for its shady practices, incompetence and dishonesty in its gross exaggeration of the influence of human CO_2 emissions on the climate. He wrote a book entitled, *The Cold Sun: Why the Climate Disaster Won't Happen.*

---

111 Dr. Tim Ball, "A History of Dr. Ben Santer and His IPCC 'Trick'", WattsUpWithThat.com, Jan 24, 2019

112 Larry Bell, op. cit.

In 1993 US Senator Tim Wirth (D-CO), who was on board with all of this, revealed the true motivation for this deception. "We've got to ride the global warming issue. Even if the theory of global warming is wrong, we will be doing the right thing."

Wirth had arranged for NASA's John Hansen to deliver testimony before Congress about global warming on June 23, 1988, a day they expected to be one of the hottest of the year. Later he candidly admitted on an *ABC* "Frontline" broadcast, "We went in the night before and opened all the windows, I will admit, right, so that the air conditioning wasn't working inside the room...it was really hot."[113]

Sometimes the perpetrators of a deception put the truth right out in plain sight. In 1998 Christine Stewart, Canadian Minister of the Environment, commented, "No matter if the science is all phony, there are collateral benefits...Climate change [provides] the greatest chance to bring about justice and equality in the world."[114]

Despite this promise of "justice and equality," the fulfillment of the IPCC goals by means of the Green New Deal or a similar plan would actually have the opposite effect. The IPCC's stated aim is to redistribute wealth from the rich developed nations to the poor third-world nations. Looking more closely, this plan would really transfer funds from the poor of the rich nations to the corrupt rulers of the third world nations, many of whom have already personally profited at the expense of their citizens.

The "carbon" taxes that have been proposed for the industrialized nations would use up a much larger proportion of the energy budgets of their poor than of the rich. The poor must still put fuel in their cars to drive to work. The poor pay a much higher percentage of their incomes to heat their homes than do the rich.

In introducing Agenda 21 Maurice Strong boldly declared, "Current lifestyles and consumption patterns of the affluent middle class - involving meat intake, use of fossil fuels,

113 Steven Goddard, "Scientists Explain Why Hansen/Wirth Sabotaged The Air Conditioning In Congress In June, 1988," https://stevengoddard.wordpress.com/2011/04/12/scientists-explain-why-hansenwirth-sabotaged-the-air-conditioning-in-congress-in-june-1988/, Aptil 12, 2011

114 Zubrin, op. cit.

appliances, air-conditioning, and suburban housing - are not sustainable." So here it was being dictated to the world's people what they'd be allowed to do.

In support of the plan, David Brower of the Sierra Club, an NGO [non-governmental organization] supporting the project, added, "Childbearing should be a punishable crime against society, unless the parents hold a government license."[115]

The goal of Agenda 21 was summarized as follows:

> "Effective execution of Agenda 21 will require a profound reorientation of all human society, unlike anything the world has ever experienced. It requires a major shift in the priorities of both governments and individuals, and an unprecedented redeployment of human and financial resources. This shift will demand that a concern for the environmental consequences of every human action be integrated into individual and collective decision-making at every level."[116]

In 1992 Agenda 21 slipped into the United States under the radar. Because it was supposedly a "voluntary" plan, the American people were never consulted or even informed of this trickery. In 1993 the Clinton Administration wasted no time in establishing the "President's Council for Sustainable Development" to begin implementing some of its goals.[117] Congresswoman Nancy Pelosi introduced a resolution of support for the plan into Congress, calling it a "comprehensive blueprint for the reorganization of human society."[118] The motivation for preparing the people for Agenda 21 - which is now posing as the Green New Deal - is "human-caused global warming".

"Social justice" has been the byword invented and used by the power elite for grabbing the moral high ground. At the same

115 Tom DeWeese, "The Principles of Freedom Vs Agenda 21," http://www.newswithviews.com/DeWeese/tom247.htm, April 22, 2014

116 Tom DeWeese, "Green New Deal Reveals the Naked Truth of Agenda 21," https://americanpolicy.org/2019/02/25/green-new-deal-reveals-the-naked-truth-of-agenda-21/, Feb. 25, 2019

117 DeWeese, op. cit.

118 Ibid.

time the World Bank has denied loans to African countries such as Nigeria to install coal fired power plants even though Africa has abundant coal reserves. If they start burning coal to produce electricity, Nigeria's thirty million people will lose the aid they receive from the European Union. Before receiving their grants, the leaders of these nations must promise to install only "alternative" energy that is impractical and 2-5 times as expensive (see chapters 13 and 14). Is this social justice?

As a result, the over 1.2 billion people who still don't have electricity will have to pay an exorbitant price for it, putting that luxury out of reach of the poor. These people are being denied the same opportunity to pull themselves out of poverty which the people of the developed countries have had. The price of food has skyrocketed, because so many food crops have been diverted to the production of ethanol and other biofuels. This unfortunate misappropriation of agricultural land caused food riots in Egypt, directly causing "Arab Spring." Is this social justice?

If the cooling trend we're currently experiencing continues, millions of people will starve due to reduced agricultural production or freeze to death if they can't afford the energy to heat their homes.

7. Manipulation of Climate Data

As unbelievable as it may seem, agencies in the US, Australia, New Zealand and other nations have actually *changed their own temperature records.* Many climate scientists have been surprised and shocked to see later editions of graphs by the National Oceanic and Atmospheric Administration (NOAA) that had been changed by that agency. NOAA moved older temperature records from the 1930's downward and more recent temperatures upward, so that temperatures would appear to have risen. In other words, just as in the novel *1984,* they were actually *changing history.*

This is not a conspiracy theory but an easily-provable fact, determined by comparing previously published graphs with the same graphs published later by these agencies after they had changed the data.

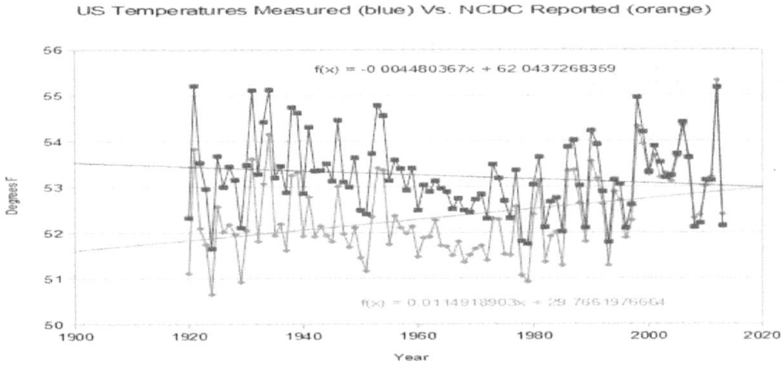

The blue line in this graph is the actual measured temperatures that were downloaded by scientists from official government websites. The red line of changed data is the graph that was ultimately released to the public. These agencies can't deny that they've cooked the data. With a flurry of double talk they claim that they've made "adjustments" to it.

Tony Heller, of RealClimateScience.com concluded that NOAA's "adjustments" represent the entire "warming trend". He says this is worse than lying with statistics. It is entirely made up.[119]

Canadian statistician Steve McIntyre caught James Hansen of NASA manipulating ("adjusting") some of his GISS data. Hansen had lowered temperatures of previous years to give the appearance that Earth had been heating up. He was forced to revise this data.[120] The same methods of fudging temperature data under the direction of the Climate Research Unit (CRU) in Britain were also used in Australia and New Zealand. Of course this action goes against every principle of science. It is blatant fraud being committed right before our eyes.

"In the year 2000 NASA showed about 0.5 warming prior to 2000. Now they show nearly 1.5C warming prior to the year 2000"[121]

Another trick that's been done to manipulate temperature records is removing some surface weather stations from the list used in compiling their averages. The stations recording colder temperatures were the ones removed from the dataset. "Over 800 worldwide ground thermometer stations have been dropped unannounced from the official global record in the space of one year despite the ongoing climate controversy. It's now been revealed that the disappearance of some ground temperature stations first began by stealth in 1990, just two years after the Intergovernmental Panel on Climate Change (IPCC) was formed by the United Nations."[122]

The EPA's National Climate Assessment has also seriously misreported data. The 2018 assessment printed graphs such as the one below: In this graph the EPA used the 1960's for its starting point, because at that time the United States had cooled from warmer temperatures of the 1930's and 1940's.

119 Tony Heller, "Climate Scientists Rewriting The Past," www.realclimatescience.com/climate-scientists-rewriting-the-past/
120 Booker, op. cit.
121 Tony Heller, "Doubling The Hockey Stick Fraud", https://realclimatescience.com/tripling-the-hockey-stick-fraud/
122 http://climateaudit.org/2007/04/27/the-team-never-makes-it-easy/

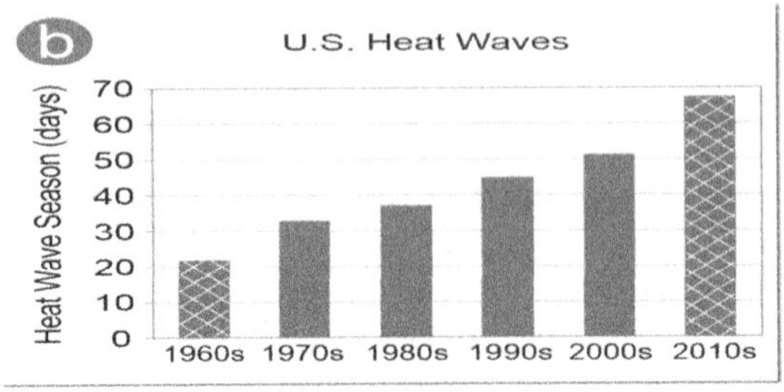

But the truth is that in the decade of the 1930's it was much warmer. In fact, thousands of people died in the 1930's heat waves. During that decade it was hotter than the first 18 years of the 21st century, despite what you've heard or read to the contrary.

As you can see from the graph below, if the "US Heat Waves" graph shown above had been superimposed on the climate record starting with the year 1900, it would have come out looking like this. Notice how much greater the heat waves were in the 1930's and 40's.

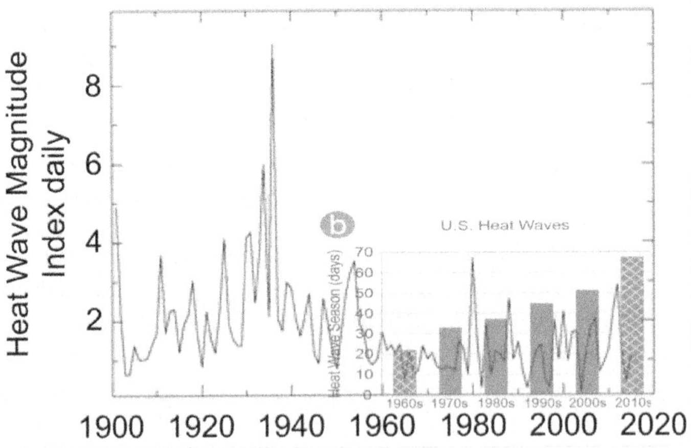

https://science2017.globalchange.gov/chapter/6/
https://nca2018.globalchange.gov/downloads/NCA4_2018_FullReport.pdf

"The 2018 report released to the public, cherry-picked the only start date in the graph which would create the appearance of a warming trend."[123] The colder decade of the 1960's was the only starting point the EPA could have used to make it look as if

123 Ibid.

the last two decades were the warmest on record. They simply acted as if the previous weather records didn't exist. This is fraud of the highest order. It should make people very angry when they realize that such fraud paid for by their taxes supports the expenditure of billions more of their tax dollars on a nonexistent problem.

In 2009 Dr. Alan Carlin, an employee of the Environmental Protection Administration (EPA) was given five days to review the Draft Technical Support Document for Endangerment Analysis for Greenhouse Gas Emissions under the Clean Air Act. In an 85-page paper he made 19 recommendations for revisions to that document. He and Dr. John Davidson criticized the EPA for accepting the IPCC's mistaken conclusions and failing to do their own scientific research. Some of their conclusions were:

> "Global temperatures have declined extending the current downtrend to years. At the same time atmospheric CO_2 levels have continued to increase and CO_2 emissions have accelerated."
>
> "The consensus on past, present and future Atlantic hurricane behavior has changed...Now the consensus is much more neutral, arguing that future Atlantic tropical cyclones will be little different than those of the past.
>
> "The idea that warming temperatures will cause Greenland to rapidly shed its ice has been greatly diminished by new results indicating little evidence for the operation of such processes."
>
> "Given the downward trend in temperatures since 1998 (which some think will continue until at least 2030), there is no particular reason to rush into decisions based upon a scientific hypothesis that does not appear to explain most of the available data."

Sen. Jim Inhofe (R-OK), one of the few political voices speaking out against climate alarmism, was overjoyed about the report, feeling that he had finally been vindicated. However, a senior EPA official, Al McGartland, quickly put the lid on it, admonishing its writers in an email, "The administrator and the

[Obama] Administration have decided to move forward [with a cap-and-trade bill]...and your comments do not help the legal or policy case for this decision."

8. "Climategate"

As if this wasn't bad enough, in 2009 an astounding fact was revealed. 1,079 emails that had been exchanged between climate scientists at the University of East Anglia's Climatic Research Unit (CRU), headed by Dr. Phil Jones, in Britain and other climate scientists with whom the CRU was working, were leaked. Other scientists involved included Keith Briffa, Ben Santer, Kevin Trenberth, Gavin Schmidt, James Hansen and Michael Mann.

Because of the nature of these emails, journalist James Delingpole coined the term "Climategate" to describe the scandal, and the name stuck. Other scientists were shocked and dismayed at what these emails revealed. This small group of scientists were those who for years had been most influential in driving the worldwide alarm over global warming and playing a major role in the conclusions of the IPCC.

One email which received a lot of notoriety indicates that Michael Mann of Penn State University and others were tampering with data. This email reads, "I've just completed Mike's Nature trick of adding in the real temperatures to each series for the last 20 years [i.e. from 1981 onwards] and from 1961 for Keith's to hide the decline."

Several of these emails centered on the controversy over a temperature graph published by Michael Mann. The statistical methods he had used to create the uptick on the right side, which to some resembled a hockey stick, were first exposed as fundamentally flawed by Steve McIntyre, a Canadian statistician. A heated battle was raging between Mann's supporters and McIntyre over the statistical basis on which the CRU had constructed the global warming case for the IPCC. Based on data taken from tree rings, this graph showed that, after a thousand years of global temperatures being very steady, they had suddenly risen to their highest level in recorded history.

However, that isn't what the tree ring data showed. They actually indicated that temperatures had *declined* during the last twenty years of the twentieth century. This wasn't what Mike Mann had anticipated. So he decided to lop off the tree ring data from 1961 onward and *replace it with data from thermometers*. One leaked email stated, "Let's use Mike's trick to hide the decline." This data wasn't revealed for a long time, but when it was finally leaked, it turned out that the tree ring data did indeed reveal a "decline" in temperatures after 1961.[124]

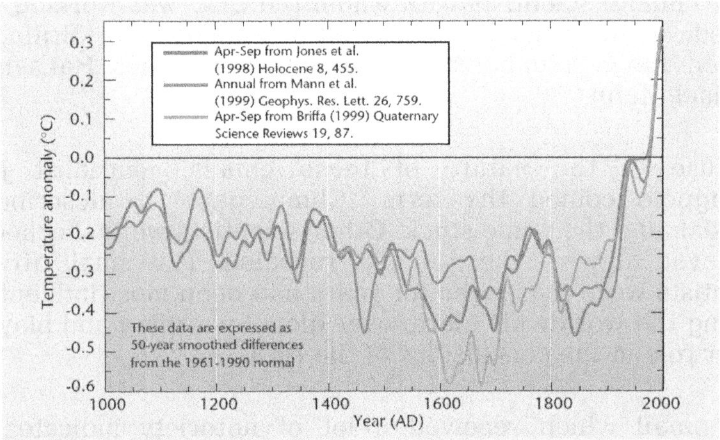

The resulting graph resembling a "hockey stick", shown above, on which the Medieval Warm Period had been obliterated, was given a great deal of publicity. It was published everywhere, including *Nature* magazine, but no one had bothered to check the data behind it. "Given star billing by the IPCC, not least for the way it appeared to eliminate the long-accepted Mediaeval Warm Period when temperatures were higher than they are today, the graph became the central icon of the entire man-made global warming movement."[125]

Mann had used tree ring data to determine what temperatures were before people were around to record them with thermometers. Data obtained from such studies are called "proxy data". But factors other than temperature, e.g. rainfall, CO_2 concentration and soil, can also determine the growth rate

---

124 "Climategate 'hide the decline' explained by Berkeley professor Richard A. Muller", YouTube Video posted on 2/27/11

125 Christopher Booker, www.telegraph.co.uk/comment/columnists/christopherbooker/6679082/Climate-change-this-is-the-worst-scientific-scandal-of-our-generation.html, Nov. 28, 2009

of trees.

The leaked email documents of "Climategate" revealed discussions between these men about how they could manipulate data to "get rid of" the Medieval Warm Period. They also revealed that this group also considered tactics to prevent the release of the data on which the IPCC's conclusions were based, which were being requested by other scientists under Freedom of Information (FOI) laws. In Britain failure to release FOI requested data is a criminal offense. Phil Jones claimed to have lost large chunks of this data.

Jones had commented,

> "I should warn you that some data we have we are not supposed to pass on to others. We can pass on the gridded data - which we do. Even if WMO agrees, I will still not pass on the data. We have 25 or so years invested in the work. Why should I make the data available to you, when your aim is to try and find something wrong with it?"[126]

This attitude flies in the face of the scientific method, an atmosphere of transparency in which scientists make requested data available to others who are endeavoring to understand how they reached their conclusions.

In these email communications the scientists also expressed their frustration at the fact that the temperatures had failed to rise as expected. If they were worried about the earth warming, they should have been happy that temperatures had leveled off.

Equally shocking, these men revealed the ruthless methods they used to silence anyone who dared to question their data and conclusions. They had pressured editors of scientific journals to block the publication of studies that contradicted their results. A professional scientist's career rests on his/her success in getting reports of their studies published in "peer reviewed" journals. "Peer-reviewed" means that other professional scientists in that branch of science affirm that the study is good enough to be published.

---

126 Email to Warwick Hughes, Feb. 21, 2005

For example, in March of 2003 Phil Jones wrote to his colleagues:

> "A paper by astrophysicists Willie Soon and Sallie Baliunas was published by *Climate Research*, which concluded that 'the 20th century is probably neither the warmest nor a uniquely extreme climatic period of the last millennium.'
>
> "Tim Osborn has just come across this. Best to ignore probably, so don't let it spoil your day. I've not looked at it yet. It results from this journal having a number of editors. The responsible one for this is a well-known skeptic in New Zealand. He has let a few papers through by [skeptics] Michaels and Gray in the past. I've had words with Hans von Storch about this, but got nowhere...
>
> "Writing this I am becoming more convinced we should do something ...I will be emailing the journal to tell them I'm having nothing more to do with it until they rid themselves of this troublesome editor. A *Climatic Research* Unit person is on the editorial board, but papers get dealt with by the editor assigned by Hans von Storch."

The response to this from Mike Mann was:

> "The Soon and Baliunas paper couldn't have cleared a "legitimate" peer review process anywhere. That leaves only one possibility—that the peer review process at *Climate Research* has been hijacked by a few skeptics on the editorial board. And it isn't just De Freitas; unfortunately, I think this group also includes a member of my own department... The skeptics appear to have staged a "coup" at *Climate Research* (it was a mediocre journal to begin with, but now it's a mediocre journal with a definite "purpose")."

Mike Mann continued:

> "I told Mike MacCracken that I believed our only

choice was to ignore this paper. They've already achieved what they wanted—the claim of a peer reviewed paper. There is nothing we can do about that now, but the last thing we want to do is bring attention to this paper, which will be ignored by the community on the whole... It is pretty clear that the skeptics here have staged a bit of a coup, even in the presence of a number of reasonable folks on the editorial board...

"So what do we do about this? I think we have to stop considering *Climate Research* as a legitimate peer-reviewed journal. Perhaps we should encourage our colleagues in the climate research community to no longer submit to, or cite papers in, this journal."[127]

In preventing scientists whose findings disagreed with theirs from being published, this small clique was not only damaging the careers of other scientists but also killing debate and preventing alternative points of view from finding their way to the United Nations decision makers. They also insulted dissenting scientists. For example, they called Piers Corbyn, an exceptionally successful meteorologist, a "prat" (British for "useless idiot").

Jones demonstrated corruption of the peer review process in several emails such as the following to Mike Mann:

"You may think Keith or I have reviewed some of your papers but we haven't. I've reviewed Ray's and Malcolm's—constructively, I hope, where I thought something could have been done better. I also know you've reviewed my paper with Gabi Hegerl very constructively."[128]

Peer reviews are supposed to be done anonymously. But members of this clique were basically agreeing to peer review one another's papers in order to expeditiously get as many as possible published in scientific journals.

In 1999 Phil Jones wrote to Mike Mann,

127 Costella, op. cit., pp. 29-30
128 Costella, op. cit., p.10

> "I must admit to having little regard for the Web. Living over here makes that easier than in the US—but I would ignore the so-called skeptics until they get to the peer-review arena. I know this is harder for you in the US and it might become harder still at your new location. I guess it shows though that what we are doing is important. The skeptics are fighting a losing battle."

Several emails indicated that they were trying to make the results of their research as sensational as possible, whether correct or not. On Sept. 19, 1996, Gary Funkhouser wrote,

> "I really wish I could be more positive about the ... material, but I swear I pulled every trick out of my sleeve trying to milk something out of that....I don't think it'd be productive to try and juggle the chronology statistics any more than I already have—they just are what they are ... I think I'll have to look for an option where I can let this little story go as it is."[129]

Geoff Jenkins' letter to Phil Jones on Nov. 22, 1996, reads in part,

> "Remember all the fun we had last year over 1995 global temperatures, with the early release of information (via Australia), 'inventing' the December monthly value, letters to *Nature*, etc., etc.?
>
> "I think we should have a cunning plan about what to do this year, simply to avoid a lot of wasted time...We feed this selectively to Nick Nuttall (who has had this in the past and seems now to expect special treatment) so that he can write an article for the silly season...I know it sounds a bit cloak-and-dagger but it's just meant to save time in the long run."[130]

After these emails were leaked, other climate scientists such as Dr. Eduardo Zorita and George Monbiot called for Dr. Jones to

---

129 https://www.lavoisier.com.au/articles/greenhouse-science/climate-change/climategate-emails.pdf

130 Ibid.

step down as head of the CRU. But he has refused to do so. Because the media covered up this affair - which should have been the biggest story of the year - most people have no idea that this important information ever surfaced.

You may have wondered how the term "global warming" changed to "climate change". This was no accident. A 2004 email from Minns/Tyndall Center on the UEA campus stated, "In my experience with global warming freezing is already a bit of a public relations problem with the media."

To this came the reply from Bo Kjellen in Sweden: "I agree with Nick that climate change might be a better labeling than global warming."

A letter to the Norwich Green Party by Ian Harris, who worked for CRU, reveals the desire these scientists had to use scare tactics to force public policy:

> "We're looking at an unprecedented acceleration in temperature ... Even if it turns out to be naturally-occurring, who's willing to take that chance? We should be trying to wean ourselves off of unsustainable energy generation and use anyway."

Harris was saying in effect that even if the warming should later be found to be from natural causes rather than from the actions of mankind, he didn't care.[131]

Christopher Booker has called this "the greatest scientific scandal of our age."[132]

Other evidence of corruption within the climate alarmist circle has also surfaced. Testifying before a congressional committee on Dec. 6, 2006, Dr. David Deming, a University of Oklahoma professor in the College of Earth and Energy, stated, "I received an astonishing email from a major researcher in the area of climate change. He said, 'We have to get rid of the Medieval

---

131 Costella, op. cit., p.26

132 Christopher Booker, *"The Real Global Warming Disaster: Is the Obsession with 'Climate Change' Turning Out to be the Most Costly Scientific Blunder in History?", 2009*

Warm Period.'"[133]

Two years later, in 2011, 5000 more emails were leaked, in what is now called Climategate II. These were even more startling and incriminating than the 1079 that had been released in 2009. Myron Ebell commented,

> "If there were any doubts remaining after reading the first Climategate emails, the new batch of emails that appeared on the web today [November 22] make it clear that the UN Intergovernmental Panel on Climate Change is an organized conspiracy dedicated to tricking the world into believing that global warming is a crisis that requires a drastic response."[134]

In these emails Phil Jones, head of the CRU, indicated his continued efforts to keep important data under wraps.

> "I've been told that IPCC is above national FOI Acts. One way to cover yourself and all those working on AR5 would be to delete all emails at the end of the process. Any work we have done in the past is done on the back of the research grants we get - and has to be well hidden. I've discussed this with the main funder [U.S. Department of Energy] in the past and they are happy about not releasing the original station data."

He also indicated that there were serious credibility problems with the models they were using.

> "Basic problem is that all models are wrong - not got enough middle and low level clouds. ...what he [Zwiers] has done comes to a different conclusion than Caspar and Gene! I reckon this can be saved by careful wording."

---

133 Tony Heller, "Erasing The Medieval Warm Period" https://realclimatescience.com/2018/12/erasing-the-medieval-warm-period/, Dec. 23, 2018

134 Myron Ebell, Director of the Competitive Enterprise Institute's Center on Energy and Environment.

Another scientist truthfully pointed out,

> "It is inconceivable that policymakers will be willing to make billion-and trillion-dollar decisions for adaptation to the projected regional climate change based on models that do not even describe and simulate the processes that are the building blocks of climate variability."[135]

You might think that, with these shady practices and double dealing revealed, the public would have demanded an end to this hoax. But again, this scandal carefully disappeared from view as if it had never happened. Everything was carefully covered up, and most of the public still has no idea that any of these shenanigans had been revealed.

135 Larry Bell, "Climategate II: More Smoking Guns From The Global Warming Establishment," www.forbes.com, Nov. 29, 2011

9. Natural Causes of Climate Variation

Scientists who wanted to profit from lucrative research grants followed the lead of the IPCC in considering only human generated causes of warming: mainly the production of carbon dioxide from burning fossil fuels.

Most scientists agree that as insignificant as the percentage of carbon dioxide is compared to water vapor, its increase in the atmosphere may have caused a *slight* amount of warming. Whereas atmospheric carbon dioxide now stands at 400 ppm, water vapor is 30,000 ppm. CO_2 controls only about 1.5% of all "greenhouse" warming.

Dr. Patrick Moore, co-founder and former CEO of Greenpeace, has drawn a remarkable conclusion.

> "Human emissions of carbon dioxide have saved life on Earth from inevitable starvation and extinction due to lack of CO_2. To use the analogy of the Atomic Clock, if the Earth were 24 hours old, we were at 38 seconds to midnight when we reversed the trend towards the End Times. If that isn't good news I don't know what is. You don't get to stave off Armageddon every day.
>
> "I issue a challenge to anyone to provide a compelling argument that counters my analysis of the historical record and the prediction of CO_2 starvation based on the 150 million year trend...I submit that much of society has been collectively misled into believing that global CO_2 and temperature are too high when the opposite is true for both. Does anyone deny that below 150 ppm CO_2 plants will die? Does anyone deny that the Earth has been in a 50-million-year cooling period and that this Pleistocene Ice Age is one of the coldest periods in the history of the planet?"[136]

136 Patrick Moore, "Shouldn't We Celebrate Carbon Dioxide?"

The 3.3°C increase in temperature the computer models predicted for a doubling of carbon dioxide has been proved wrong by the last twenty years of temperature readings. These predictions simply did not materialize. While the amount of carbon dioxide has continued to increase in the past 19 years, the temperature has been stable. During this time 30% of mankind's contribution of CO_2 to the atmosphere since the early 1900's has occurred. Most of the increase in carbon dioxide is from natural causes. In fact, only about 3.4% of the increased CO_2 in the atmosphere is the result of human emissions.[137]

Several scientific studies, notably by Drs. Will Happer, Richard Lindzen, and Roy Spencer have calculated the maximum temperature change that could be expected for a doubling of CO_2. All of these studies reached the same figure: an increase of 0.6°C (±0.1°C). In contrast, the models predicted an increase of 3.5°C (±1.5°C, meaning anywhere from 2°C to 4.5°C).

One reason for the inaccuracy of the models' predictions, as explained previously, is that they failed to factor in the buffering effect, or negative feedback, of cloud cover caused by increased humidity.

Some warming may have been caused by changes in land use, especially the destruction of the rain forests and urbanization. The average temperatures of the earth's great cities, such as New York, have risen significantly in the past century. Yet the average temperatures of many smaller cities has been flat or even reduced during the same period of time.

However, the United Nations commissioned the IPCC to study average temperature increases due ***only*** to the carbon dioxide emitted by human activities. They made no allowance for any natural cause of climate change. This leaves out all natural factors known to influence climate, such as cyclical changes in solar activity, changes in the earth's orbit, caused by the gravitational pull of Jupiter, the tilt of Earth's axis of rotation and the precession of the equinoxes.

Since the 1990's Dr. Henrik Svensmark and his team have been working on an alternate hypothesis. This ongoing study is based

---

http://www.freerepublic.com/focus/f-news/3730010/posts?page=13, Oct.15, 2015

137 Ball, op. cit., p. 27

on the well known fact that the sun is one driver of changes in the climate. Sunspots - some the size of the earth - are huge magnetic storms on the sun's surface. The sun's magnetic field is stronger during the peaks in its 11-year cycles of sunspot activity. When there are more of these magnetic storms on the sun, more tiny charged particles, called the "solar wind", are released into space. A number of these particles reach Earth. This solar wind, working with Earth's magnetic field, deflects cosmic rays heading toward Earth from outer space.

Cosmic rays are not rays at all but highly energized atomic-sized particles entering Earth's atmosphere at nearly the speed of light. These are believed to originate from huge explosions called supernovae, in which massive stars in our galaxy end their lives.

Svensmark and his colleagues hypothesized that these cosmic rays interact with organic molecules produced by trees to provide particles called condensation nuclei around which water vapor condenses to form the water droplets making up clouds.

There are more cosmic rays in the atmosphere during the low part of the sun's 11-year cycle when sunspots are at a minimum.

There is an extremely close correlation between the percentage of low cloud cover and the number of cosmic rays reaching a point on Earth at any given time. It is true that correlation does not equal causation. However, during the 22-year period between 1983 and 2005, this correlation was 92%: extremely close. And this has been the case during the 400-year history of sunspot observation. So when there are more cosmic rays reaching Earth, there is also a higher percentage of low clouds, which reflect sunlight and cool things off.[138]

CERN laboratory in Switzerland, which studies collisions between atomic-sized particles traveling at nearly the speed of light, tested this theory.

"In the artificial atmosphere, scientists directed a

---

138 Larry Vardiman, A New Theory of Climate Change, 11/1/08 https://www.icr.org,

> beam of positively charged particles, so-called pions, which were generated by a Proton-Synchrotron. With this they simulated the effect of cosmic rays, which are constantly penetrating into the atmosphere. This stream of particles caused the organic molecules to ball up into approximately 2-nanometer size particles. More measurements confirmed that these volatile organic compounds grew further to form efficient condensation nuclei. In parallel the scientists underpinned the mechanism with computer models. 'These results for me are the most important that the CLOUD experiment yielded so far,'"[139]

One fact that makes this theory so attractive is the extremely close correlation between the degree of sunspot activity and average temperatures on Earth. You will never hear this from the media. Another fact the major media haven't reported is that when average temperatures on Earth were cooler, as they were in the 1970's, other planets in the Solar System, such as Venus and Mars, also had lower average temperatures.

Not all solar cycles are the same. In strong cycles there are many large sunspots, and in others hardly any. At the end of solar cycle #24, the sun was in a period of very low solar activity with absolutely no sunspots over a period of several days. A website on which you can see what the sun looks like on any given day is SpaceWeather.com.

The solar minimum between solar cycle #23 and #24 saw 260 spotless days in 2009 - the most days without sunspots in a single year since 1913.[140] This is very unusual. During the Little Ice Age, a prolonged period of low sunspot activity known as the Maunder Minimum preceded an era so cold that between 1645 and 1715 even the Thames River in London froze up in the winter, and people regularly ice skated on it.[141]

As solar cycle #24 is drawing to a close, many scientists are predicting that cycle #25 will have even fewer sunspots. Dr. Fred Goldberg predicted,

---

139 Lofken, Wissenschaft, May 26, 2016, translation in notrickszone.com

140 http://www.climatedepot.com/2018/05/23/are-we-headed-towards-another-deep-solar-minimum/

141 Mark Prigg, "Is a Mini Ice Age On the Way?", DailyMail.uk, July 10, 2015

> "We have now just started a climate journey into a new little ice age, which will last 20 to 30 years. This will lead to less food production, increased food prices, famine disasters, higher energy consumption, higher energy prices, some very cold winters good for ice skating, more traffic accidents, etc."[142]

The sun goes through other cycles besides the 11-year one. Otto-Carl Weiss and his colleagues examined European temperature data for the past two hundred fifty years using Fourier Transform spectral analysis. Their measurements indicated that, rather than a constant increase in temperature during the time period since the start of the Industrial Revolution, the climate had warmed and cooled periodically in ~200-year De Vries solar cycles. Increased warming for the last two decades of the 20th century and cooling for the previous two decades are due to 65 years of changes in ocean currents. The team also concluded that climate change is natural with no influence from carbon dioxide.[143]

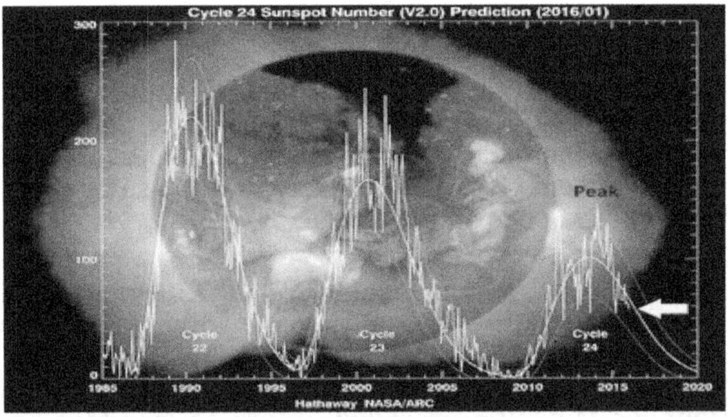

This graph of solar cycles visually represents the average number of sunspots during the last three eleven-year cycles. It is evident that the sun was much more active during the last two decades of the 20th century. During the first two decades of the 21st century there were fewer magnetic storms on the sun. Some scientists are predicting that solar cycle #25 will have the lowest number of sunspots of any cycle in the past 200 years.

142 Lecture by Dr. Fred Goldberg, March 15, 2016

143 Lecture by Prof. Carl-Otto Weiss, Advisor to the European Institute for Climate and Energy, "International Conference of the Schiller Institute", Paris, France; June 14, 2015

A new theory of the natural cause of climate change will soon be introduced by Dr. Don Easterbrook, who conducted an extensive investigation of all warm/cold periods, not just full-scale Ice Ages and interglacial warm periods but also short-term warming and cooling periods. He states:

> "For each warm/cold period, I looked at temperature, sunspot numbers (SSN), total solar irradiance (TSI), and radiocarbon (14C) and beryllium-10 production rates as indicators of cosmic ray intensity. The data are truly remarkable. Every cold period, regardless of the magnitude of the warm or cooling, was characterized by low sunspot numbers, low TSI, and high production rates of 14C and 10Be, indicating high cosmic ray intensity. Every warm period was characterized by high sunspot numbers, high TSI, and low production rates of 14C and 10Be, indicating low cosmic ray intensity."

The results Dr. Easterbrook obtained from this study are beyond any possibility of coincidence. Therefore, they would have to be causes and effects. This study supports the conclusions of the Svensmark Clouds study, but does not depend on them. It stands alone on its own merits. The results of this study will soon be published as a book "Solar Magnet Cause of Climate Changes and the Origin of the Ice Ages".[144]

Patterns of ocean currents are a large factor in minor warming and cooling of the climate. The Pacific Decadal Oscillation (PDO) is one such pattern. It reverses approximately every sixty years. When warmer water from the south Pacific moves through the Bering Strait into the Arctic Ocean, the Arctic is warmed, and ice melts. The PDO switched to this pattern in 1910, after which warming occurred, and again in 1977, after which the warming pattern repeated. It is interesting how these two dates line up with the beginning of each warming period occurring in the 20th-century.[145]

Climate changes occurring over much longer time periods are due to cyclical variations in Earth's orbit. For example, the earth's axis tilts anywhere from 21° to 24.5° to a line perpendicular to the plane of its orbit around the sun. Currently

144 Dr. Don Easterbrook, from abstract, July 9, 2019
145 Dr. Fred Goldberg lecture March 15, 2016 (YouTube video)

that angle is about 23°. Earth wouldn't have seasons if its axis weren't tilted with respect to the plane of its orbit around the sun. The earth's axis also wobbles like a top. Milutin Milankovitch (1879-1958) correlated these changes with ice ages that have been occurring on a regular basis throughout geologic history. So climate change occurs in cycles.

10. Following the Money

If human-caused CO_2 emissions - responsible for about 3.4% of the trace gas carbon dioxide - aren't significantly warming the earth, then it follows that the trillions of dollars already spent or planned to reduce these emissions would make little or no difference in the resulting temperature. The Small Business Administration has estimated that compliance with all of the regulations - many of which are state regulations - already in place is costing the US economy over $1.75 trillion per year — about 12%-14% of the US Gross Domestic Product (GDP). This is about half of the $3.5 trillion per year that the federal government currently spends.[146]

These wasted trillions of dollars could have been spent on solving the earth's *real* environmental problems, such as recycling plastics, metals and paper, deforestation, pesticides, nuclear radiation, dioxin and nitrates in our water supplies, to name a few. Just the $3.4 billion spent each year to pump carbon dioxide into the ground [sequestration] could be used to provide clean water each year to a hundred million people who don't now have it.[147] This would save millions of lives.

"The US government has spent over $79 billion since 1989 on policies related to climate change, including science and technology research, administration, education campaigns, foreign aid, and tax breaks."[148] Scientists have been handsomely paid to find a human-generated crisis. Remember that if they stop doing this, their grants will be discontinued. In other words, they'll be out of a job. By the year 2009, scientists investigating human caused CO_2 emissions had already been awarded $32.5 billion.

So when you begin to follow the money, the reason for the thousands of articles and news reports sounding the alarm becomes evident. Controlled by just six huge corporations which represent those in power, the obedient media incessantly grinds out articles and news stories all over the world about the

146 Henry W. Burke, "Global Warming: an Expensive Myth, Pt. 2," Feb. 24, 2013
147 Joanne Nova, *The Skeptic's Handbook*, 2009
148 Joanne Nova, *Climate Money*, 2009

harmful effects of "climate change".

At first, in the late 20th century, they said that snowfall would diminish. Then, when there was more snow than ever, they told us that this was due to "climate change". Carbon dioxide produced by mankind has been blamed for more lightning strikes and also for fewer lightning strikes, for higher temperatures and also for lower temperatures. So whatever happens with weather events, they are right. Essentially this makes it impossible to disprove the "theory". But, according to the famous philosopher of science, Karl Popper (1902-1994), if a hypothesis can't be disproved, then it isn't a valid hypothesis. It isn't even science, but a political movement that has hijacked both environmentalism ***and*** science. In fact, Professor Murry Salby noted, "Climate projections are a platform for social advocacy."[149]

One goal of those who are behind this political movement is the institution of carbon trading, which has yielded billions of dollars for well placed individuals and corporations. In this scheme, also called emissions trading, or cap and trade, businesses are forced by governments to trade their "pollution rights" through a complicated system of regulations. The companies which produce less CO_2 can sell their unused "pollution rights" to companies that produce more carbon dioxide. This, of course, provides a financial incentive for companies to produce less CO_2. But it is the consumers and taxpayers who ultimately pay the bill in the form of higher prices and taxes.

By the year 2008 $126 billion had already been paid out in these cap-and-trade schemes. This practice alone is expected to generate 2 to 10 *trillion* dollars. It was engineered by the bankers of such firms as J P Morgan Stanley, Barclays, Citigroup, and Credit Suisse, who, in brokering these deals, are siphoning money away from hard working consumers and taxpayers. Casey Mulligen, reporting in the *New York Times,* stated that in the United States, "Some estimates suggest that the amount at stake for the taxpayers is [annually] over $3,000 per person."[150]

149 Murry Salby, op. cit.
150 https://economix.blogs.nytimes.com/2009/05/27/how-much-will-cap-and-trade-cost/

So the hypothesis of ***human-caused*** climate change, which has never been scientifically verified and can't be disproved, has reaped huge profits for bankers, solar and wind corporations and companies set up to benefit from cap and trade. It has also provided careers for a new generation of scientists with a vested interest in this "problem".

As will be explained in subsequent chapters, wind and solar energy will not solve the "problem", because these highly acclaimed but intermittent energy sources must be continually backed up by traditional power sources at about 83% of the power rating of the alternate fuel source. So, a large wind turbine rated for 1.5 megawatts, even under ideal conditions, needs a traditional power station on standby that is capable of feeding 1.30 megawatts of power into the grid.

Electric cars do nothing to mitigate the situation. Their only benefit is to improve air quality in cities. This is because the batteries providing the power for these vehicles are usually charged by power from the traditional hydrocarbon energy sources. Charging a vehicle's battery with energy from wind or solar is several times as expensive as with traditional sources.

Summary Points

"Much of what the media and activist groups tell us about energy is misleading or entirely false, which can lead to dangerous and expensive outcomes."[151]

As this book has already demonstrated:

No one will deny that the average temperature of Earth has risen slightly since the start of the Industrial Revolution around 1850. Since the late 1700's the earth has been coming out of the Little Ice Age. The warming experienced during the final two decades of the 20th century was totally within the bounds of natural warming. During the two decades previous to that (from 1955 to 1978) the earth had been *cooling* while the rate of carbon dioxide was rising at a steady rate. During the first two decades of the 21st century the earth's average temperature has remained fairly constant even though the percentage of atmospheric carbon dioxide has continued to rise at the same rate as during the second half of the twentieth century.

1.) The entire global warming hypothesis is based on the unscientific methodology of computer modeling, which is only as good as the information put into the computers. This is not true empirical science, which involves observation and measurement of the real world. These models predicted a warming of at least three times that which actually occurred in the 40 years since 1979. Science should be able to make valid predictions. The failure to do so indicates an invalid hypothesis that should have been scrapped.

Yet, rather than admitting they were wrong, those promoting this false hypothesis have doubled down and issued even more extreme warnings. The 6th IPCC report issued in 2018 gave us a tipping point of 12 years that we have to "fix" the problem. Why should we believe that prediction when none of the previous

151 Quote from web page of The Clear Energy Alliance.

predictions have been correct? And why are governments basing expensive policies on a failed science issuing failed predictions?

To date there has never been a single study that was able to prove that human-produced carbon dioxide is the cause of any warming we've experienced.

2.) From studying ice core samples going back 150,000 years in Earth's history, it's been known since 1999 that temperature increases have *always* occurred ***before*** the increase in carbon dioxide levels. Sometimes there is a lag of about 800 years between the temperature increase and the increase in carbon dioxide. But the lag time is sometimes as short as six years. By 2003 all climatologists were aware of this fact.

However, we were told in Al Gore's documentary, *An Inconvenient Truth,* and by incessant media reports that the opposite is the case - that an increase in temperature occurs after the carbon dioxide level rises. Gore's advisers who produced this film must have known the truth. Therefore, the statement that an increase in CO_2 causes a temperature increase is a deliberate lie.

In this 2006 documentary, for which Gore received an Oscar and a Nobel prize, he said,

> "The planet has a fever. If your baby has a fever, you go to the doctor. If the doctor says you need to intervene here, you don't say, 'Well, I read a science fiction novel that told me it's not a problem.' If the crib's on fire, you don't speculate that the baby is flame retardant. You take action."

He warned that we had just ten years left to do something.

3.) In studying the Earth's geologic history, we know that the climate has always been changing without mankind's help. As recently as 15,000 years ago the Earth was coming out of an ice age in which the area that is now New York city was covered by a huge glacier. There have been several such ice ages in geologic history. We are presently living in an interglacial period, and the next full-blown ice age could be upon us in just a few thousand years.

4.) It is now recognized that the sun, which goes through regular cycles, is one of the main drivers of climate change. Other natural mechanisms of climate change, such as cosmic rays from our Milky Way galaxy, which provide condensation nuclei that enable clouds to form, are now beginning to be understood.

5.) The earth's average temperature is presently about 15°C. The average temperature of Earth during an ice age ranges from 10°C to 12°C., just a few degrees below where it is now. It's been estimated that cold weather kills nine times as many people as warm weather. The warmer periods in history – such as the Roman warm period and the Medieval warm period, in which Earth's temperature was higher than today – were accompanied by prosperity and a flowering of civilization.

6.) Water vapor is the most important greenhouse gas in the atmosphere. Carbon dioxide is a trace gas, only about 0.04% (400 ppm) of the atmosphere. Currently its percentage of the atmosphere is one of the lowest levels at which it has ever been. Carbon dioxide is not a pollutant but food for plants. All life on Earth depends on this gas. Since the percentage of CO_2 in the atmosphere has increased slightly, forests and all other plants are growing faster and using water more efficiently. Agricultural production has increased at least 15% and thus has been able to feed more people. Should the level of CO_2 ever drop below 150 ppm, just about all life on earth would die.[152]

7.) Despite a long search taking place over many years, no global warming "signature" - or "hot spot" in the mid troposphere - at an altitude of about five km – in the tropics has ever been found. This is one of the strongest pieces of evidence we have against the human-caused global warming hypothesis.[153]

8.) Climate change alarmists have repeatedly shown themselves

---

152 Dr. Patrick Moore, "Patrick Moore: Shouldn't We Celebrate Carbon Dioxide", http://www.freerepublic.com/focus/f-news/3730010/posts?page=13, Oct. 15, 2015

153 Roy W. Spencer, PhD, "New Satellite Upper Troposphere Product: Still No Tropical 'Hotspot'",http://www.drroyspencer.com/2015/05/new-satellite-upper-troposphere-product-still-no-tropical-hotspot/, May 21, 2015

to be dishonest. First, they've floated many myths (see chapter 5) to scare people. The scare stories presented by the media reek with dishonesty.

Second, the Climategate scandals involving over 6,000 leaked emails between members of the clique of climate alarmists revealed just how they had taken over the science peer review process, demonizing dissenting scientists and using tricks to manipulate data. The emails showed that they were also unwilling to share the data on which they based their conclusions.

Third, government agencies such as NOAA and NASA have been caught red-handed changing data to make it appear that the Earth has warmed more than it actually has.

It takes just ***one lie*** to discredit any hypothesis. The global warming alarmists have been guilty of many, many lies.

9.) None of the many predictions of impending doom and tipping points made during the last four decades *has ever materialized*.

10.) If the IPCC ever admitted that they'd been wrong, they'd go out of business. The pseudoscientists who have so far received over eighty billion dollars in grants would have to find other work.

Global warming scare tactics have advanced the IPCC's political objective of worldwide totalitarian control. The founder of that organization, Maurice Strong, was quoted earlier in this book as saying, "Isn't the only hope for the planet that the industrialized civilizations collapse? Isn't it our responsibility to bring that about?" Bringing down our industrial civilizations and their governments would be swiftly followed by a totalitarian power grab.

11.) An entire industry involving carbon trading and taxation and the so-called "green" alternatives of energy production, have already cost the world's people *trillions* of dollars, and this is just the beginning. By 2015 almost two trillion dollars had already been spent on "renewable" energy. And they are just getting started. Many fortunes have already been made on this industry. Al Gore, who started off as governor of Tennessee, is reportedly worth $400,000,000. Tragically, these wasted resources could have been used to solve *real* environmental problems.

12.) Because the global warming alarmists have moved the goal post by changing their cause from "global warming" to "climate change" to "extreme weather", and now even blame human-caused climate change for severe cold and snowstorms, it is now impossible to disprove this hypothesis. Whatever happens, they are right. Whether there is a flood or a drought, a heat wave or freezing cold, the cause is "climate change," caused by humans. Thus, it can't be disproved.

Karl Popper (1902-1994), defined science as follows: "If a theory can be falsified, it is science. Otherwise it is pseudoscience."[154] Any hypothesis that cannot be disproved is invalid. This is just another fact pointing to the unscientific nature of the AGW hypothesis.

13.) The burden of proof really rests on the proponents of the human-caused global warming hypothesis, because they made this claim in the first place. They have never been able to provide even one piece of real scientific evidence connecting the trace gas carbon dioxide with the 20th century warming of our planet. Meteorologist Piers Corbyn stated, "There are no scientists in the world who can produce such observational data, and we challenge anyone reading this to send us such observational evidence from the real world or find someone who can and get them to produce it."[155]

---

154 Editors of *Encyclopedia Britannica,* "Karl Popper, British Philosopher,"https://www.britannica. com/biography/Karl-Popper

155 "Climate Change Challenge with Piers Corbyn," YouTube Video, Dec. 12, 2018

11.The Abiotic Theory of Petroleum

Science textbooks tell us that oil comes from the remains of plants and animals that lived and died millions of years ago. However, there is a very good chance that this theory is incorrect and that, instead, natural gas as well as petroleum, is seeping up from Earth's mantle, the layer that lies directly beneath the crust. This theory, known as the abiotic (or abiogenic) theory of petroleum formation, states that carbon found in the melted rock [magma] below the crust has reacted with hydrogen to form methane [natural gas] as well as a number of other hydrocarbons. These reactions are rather complicated, having several intermediate stages. Granite and other silicon based rocks may act as catalysts, speeding up the reactions.[156]

This process has been demonstrated in laboratory experiments both in Russia and the United States. Under extreme conditions of heat and pressure, iron oxide, calcium carbonate and water can be converted into methane. Hydrocarbons containing up to ten carbon atoms in one molecule have been produced in this manner.[157]

There are several reasons why some scientists subscribe to the abiotic theory:

1.) As time goes by it is becoming increasingly evident that there just aren't enough fossil remains to have provided the amount of petroleum that has been discovered and used. And yet more and more oil deposits are being discovered every day. Enormous amounts of petroleum have recently been found in supergiant oil fields. An estimated 65,000 oil fields exist around the planet, of which over 900 contain at least five hundred million barrels (about 79,000,000 cubic meters) of available petroleum or the equivalent in natural gas. Some individual fields contain five billion barrels each.[158] Many of these are found in the Middle East. Russia, the United States and several other countries also have extensive deposits of hydrocarbons.

---

156 http://www.viewzone.com/abioticoilx.html
157 Ibid.
158 https://en.wikipedia.org/wiki/Giant_oil_and_gas_fields

Back in the 1970's it was widely reported that there was enough oil to last the world only another twenty years or so. In that case the world would have run out of oil in the 1990's.

2.) Geologists in Russia and the Ukraine contend that the formation of oil deposits requires pressures higher than those found in deposits of organic materials. These could be found only in the mantle, where matter is subjected to tremendous pressure.[159] Under the direction of these Russian geologists, oil rigs have been successful in drilling in basement rock in the Dneiper-Donets Basin in the Ukraine. According to Professor Vladilen A. Krayushkin, Chairman of the Department of Petroleum Exploration, Institute of Geological Sciences, Ukrainian Academy of Sciences in Kiev:

> "The drilling which resulted in these discoveries was extended purposely deep into the crystalline [non-sedimentary] basement rock, and it is in that basement where the greatest part of the reserves exist. These reserves amount to at least 8,200 million metric tons [65 billion barrels] of recoverable oil and 100 billion cubic meters of recoverable gas, and are thereby comparable to those of the North Slope of Alaska." [160]

3.) In his book, *The Deep, Hot Biosphere,* Cornell University astronomy professor Thomas Gold pointed out that deposits of petroleum almost always contain helium gas, which is not produced by living organisms. On that basis he was able to persuade the Swedish State Power Board to drill for oil in a rock that had been fractured by an ancient meteorite. Although this rock was not sedimentary and thus wouldn't contain the remains of plant or marine life, oil *was* found in that rock. Even though there wasn't enough of it to compete with other oil fields, it still was oil, nonetheless.

4.) Methane [natural gas] and other organic compounds not only exist but are abundant on Jupiter, Saturn and Uranus and on some of their moons as well as elsewhere in the Solar System. Gold argued that hydrocarbons existed at the time of

159 Environmental Literacy Council (enviroliteracy.org/energy/fossil-fuels/abiotic-theory/)
160 www.gasresources.net/energy_resources.htm

the Solar System's formation. Although this could be used as evidence for life on other planets, it is much more likely an indication that organic, carbon-based compounds can be produced by inorganic, nonliving processes.

According to data collected by the Cassini-Huygens spacecraft, which extensively photographed and mapped Titan, a large moon orbiting Saturn, scientists announced on February 13, 2008, that Titan contains within its polar lakes "hundreds of times more natural gas and other liquid hydrocarbons than all the known oil and natural gas reserves on Earth." The desert sand dunes along the equator, while having no open liquid, nonetheless hold more organics than all of Earth's coal reserves.[161] It has been estimated that the visible lakes and seas of Titan contain about 300 times the volume of Earth's proven oil reserves.[162]

Despite all of the above evidence, the abiotic theory of petroleum has been rejected by the great majority of geologists, who argue that the carbon found in most petroleum is the C-12, not C-13 isotope of carbon. They contend that even though petroleum has been found at depths greater than that of the deepest fossils ever found, this phenomenon could have been caused by tectonic plate motion, which can "radically reshuffle

161 "Titan Has More Oil than Earth", February 13, 2008. Retrieved 2008-02-13.

162 earthsky.org/space/a-pure-methane-sea-on-titan

rock strata, leaving younger sedimentary oil or gas-bearing rock beneath basement rock."[163]

These molecules of hydrocarbons have been gradually forced up to the surface through cracks in rocks. The organic materials found in petroleum deposits and used by the scientific majority as the basis for the fossil fuel idea can easily be explained by the fact that strange bacteria have very recently been discovered living in extreme environments similar to those in Earth's mantle. These bacteria, called, hyperthermophiles, have been found in hydrothermal vents, at the bottoms of volcanoes, and in places where living conditions were previously believed to be impossible. Gold believes that the mantle surface could mix with fossils in rock layers on its way up to the surface.

This theory is difficult to prove, because so far oil drills have been able to penetrate only about six miles down into sedimentary rock, where petroleum is found, and about two miles deep in other types of rock. The crust is much deeper than that, and the extreme heat of the mantle would melt the casings of oil wells penetrating that deep. But a recent study by the US Department of Energy and Lawrence Livermore Lab suggests that there might be huge methane (natural gas) deposits in Earth's mantle 60 to 120 miles beneath the surface.

Why is the abiotic theory important to mankind? Well, if this information turns out to be correct, it basically means that we don't have to worry about running out of oil. In other words, petroleum is renewable energy. That doesn't mean, however, that we shouldn't be looking for better alternatives. It just takes the pressure off of us to have to switch to forms of energy production that have proven to be inferior, such as wind, solar panels and biofuels before cost effective and sustainable alternate energy sources such as safe nuclear energy are commercially available.

Most everyone agrees that to reduce pollution, cleaner methods of producing energy other than burning hydrocarbons need to be found. But the fact remains that at the present time the world's people rely almost exclusively on this low cost source of energy which has brought about the superior living standards we enjoy.

---

163 Essay by Richard Heinberg, richardheinberg.com

Methods of petroleum exploration based on the abiotic theory are making it easier to find places where it is suitable to drill for oil and natural gas. This will further reduce the cost of using hydrocarbon-based fuels. The big oil companies, which are now making a tremendous profit on oil, don't want you to know that currently the cost of pumping a barrel of oil is is between five and ten cents. This doesn't include exploration or drilling costs but is still pennies on the dollar compared to the price for which each barrel of oil is traded. All energy sources impact the environment in some way. No one can deny that affordable, abundant energy from coal, natural gas and petroleum has made life for billions of people much better than it was when energy came from burning wood, human slaves and beasts of burden.

12. Energy From Coal & Natural Gas

There are about 1,000 coal-fired power generating units in the US, providing about 310 gigawatts [billion watts] of electric power. Some of the older coal plants are shutting down and being replaced by state-of-the-art combined cycle coal and natural gas generating plants.

China, which is becoming more and more industrialized, is building hundreds of new coal power plants that will be able to generate up to 259 gigawatts (GW) of electricity—with almost as great a capacity as all existing US coal fired plants.[164]

Overall about 1,600 coal plants are planned or under construction in 62 countries.

After a load of coal arrives at a power plant, usually in train cars, each car is literally turned upside down to unload it. The coal is then ground into a fine powder. This increases the surface area, allowing it to burn more quickly and efficiently. The powdered coal is blown into the combustion chamber of a boiler where it burns at a high temperature. The heat energy produced converts liquid water into steam.

The resulting high pressure steam is then passed through a turbine that contains thousands of propeller-like blades. The steam pushing on these blades rotates the turbine shaft at a high speed. A generator consisting of huge electromagnets made from thousands of carefully wound coils of copper wire is mounted at one end of a rotating shaft. This rotating magnetic field causes electrons to move through the coils, producing electricity. The steam which rotated the turbine condenses back into liquid water and is returned to the boiler to be heated once again.

Huge transformers raise the voltage produced in this manner to

164 Mimi Nguyen, "China Building New Coal Plants Equal to Entire US Capacity," https://www.theepoch times.com, Oct. 7, 2018

400,000 volts or more. When voltage increases, amperage decreases. The reduced amperage of current in the grid transporting this electricity to its destinations also reduces the power loss through the lines. When the electrical power nears the point of consumption – homes and businesses - the voltage is transformed down to a safe 110v-240v level that can be used.

Carbon dioxide isn't a pollutant. However, carbon monoxide, sulfur dioxide, mercury and arsenic are pollutants. Technologies are already being used to prevent their release from the coal, gasoline and natural gas combustion used for electrical generation.

Engineers have also developed ingenious processes for extracting every bit of heat energy from hydrocarbons. Up to 50% more electricity can be produced in what is known as a combined-cycle power plant. Whereas coal has one go around in traditional plants, modern electrical power plants use both gas turbines and steam turbines in the same operation. In a combined-cycle plant, air is compressed and mixed with fuel that is heated to a very high temperature. This hot-air/fuel mixture pushes against gas turbine blades, causing an electromagnet to spin rapidly, producing electricity. Then, instead of allowing the hot exhaust to be wasted, it is routed through pipes to a boiler that heats the water to steam. That steam then pushes on the blades of a steam turbine that spins another electromagnet, generating additional power. This system is called a Heat Recovery Steam Generator (HRSG).[165]

Several different methods of turning carbon dioxide into fuel have also been invented. Scientists at the Department of Energy Berkeley lab devised a way to react carbon dioxide with water to produce a mixture of hydrogen gas, an excellent fuel, and carbon monoxide gas. For a catalyst they use extremely thin gold foil with an even smaller layer of copper attached to it. This is a real breakthrough.[166]

At the Massachusetts Institute of Technology (MIT) it was discovered that if you heat carbon dioxide to a temperature of 990°C. and then run it through a membrane made of

165 https://www.gepower.com/resources/knowledge-base/combined-cycle-power-plant-how-it-works

166 https://futurism.com/new-nanomaterial-system-helps-create-synthetic-fuel-from-co2

lanthanum, calcium and iron oxide, some of it will be converted to carbon monoxide, which can then be changed into syngas fuels.[167] Waste heat from a traditional coal plant could provide the energy to heat the carbon dioxide. Although these reactions have so far been done only on a small scale, this technology looks promising.

The people at Soletair Project, created at Germany's Karlsruhe Institute of Technology (KIT), claim that their reactor can convert carbon dioxide into gasoline, diesel or kerosene. This, too, is currently being done only on a small scale.[168] With this technology underway it is probably only a matter of time until these fuels are commercially produced from carbon dioxide.

Until a better source of electrical generation, such as thorium molten salt nuclear reactors or nuclear fusion, is developed, coal will continue to be one of the world's primary sources of energy. The good news is that coal burning plants are cleaner and more efficient than ever before.

Under the 1972 Clean Water Act and the 1977 Surface Mining Control and Reclamation Act, the coal mining industry is required to restore the appearance of the land that has been strip mined. However, streams affected by this type of mining are adversely affected. They have up to one third less biodiversity than before.[169]

Natural gas, made up mostly of methane, is usually found in oil rich regions. It is sought after as a prime energy source, because it burns cleaner than either coal or petroleum. Because of new techniques in natural gas extraction, the US is now completely energy independent.

Through the science of seismology, geologists search for deposits of petroleum and natural gas by artificially creating surface vibrations and then recording the ways in which these vibrations are reflected back to the surface, revealing the characteristics of the rock layers beneath.

167 https://newatlas.com/carbon-dioxide-fuel-conversion-mit/52367/
168 https://newatlas.com/carbon-dioxide-fuel-pilot-plant-finland-kit-ineratec/46362/
169 www.nature.com/articles/s41893-018-0048-6

An analogy that explains how this works is that of bouncing a rubber ball. A rubber ball dropped on concrete will bounce in a much different way than a rubber ball dropped on sand. Likewise, seismic waves sent underground reflect off of layers of dense rock much differently than those directed at more porous rock layers. By doing this kind of detective work, geologists can interpret the results of such seismic data to determine the types of underground rock formations as well as their depths.[170]

170 https://www.e-education.psu.edu/

13. Wind Energy: Neither Green Nor Sustainable

You've been told repeatedly by the media – in magazines, TV, the Internet, etc. - that electricity produced by the multibillion-dollar wind industry is clean, essentially free energy - better for the environment than traditional sources such as coal and natural gas. None of these claims is true. They are based on the reduction of carbon dioxide emissions supposedly contributing to global warming.

Yet, such wind installations injure, maim, and kill hundreds of thousands of birds and bats each year in clear violation of federal law. Any marginal reduction in CO_2 emissions comes at the expense of these birds, including protected bird species, such as bald and golden eagles.[171]

For example, the Golden Gate Audubon Society in California reported that the wind farm at Altamont was killing 10,000 birds each year, including 1123 birds of prey. Due to a lawsuit, the Altamont wind turbines have had to cease operation for the four months annually that these birds are migrating.[172]

Offshore wind turbines have similar impacts on marine birds, and, according to the Union of Concerned Scientists, wind farms located offshore also impact fish and other marine wildlife.[173] The Daily Mirror (U.K.) reported on June 6, 2005, that scores of baby seals on Scroby Sands off Great Yarmouth were found dead -- born dead or abandoned by their mothers. Staff at the wildlife hospital involved were certain that the wind facility there was to blame.

A story the wind energy promoters don't want you to hear concerns the way in which wind turbines are constructed. Modern wind turbines depend on rare earth minerals mined

171 "Top 10 Myths About Wind Energy and Birds", Michael Hutchins, https://www.wind-watch.org/documents/category/issues/impacts/wildlife/

172 Jonathan Benson, "'Green' Debacle, NaturalNews.com, Nov 24, 2011

173 /www.ucsusa.org/clean-energy/renewable-energy/environmental-impacts-wind-power#.WqAYMM9x0xA

primarily in China.[174] Federal environmental regulations in the United States restrict rare earth mining. China has a poor record of environmental stewardship. So the process of extracting these minerals imposes wretched environmental and public health conditions on people in China's cities and towns.

It takes a tremendous amount of material to produce one wind turbine. A typical wind turbine, over 500 ft. tall, contains more than 8,000 different parts, many of which are made from steel, cast iron, and concrete. One such component is magnets made from neodymium and dysprosium, rare earth minerals mined almost exclusively in China.

Simon Parry, a reporter for the Daily Mail in Britain, traveled to Baotou, China, to investigate the mines, factories, and toxic waste dumping grounds associated with this industry. In the village of Dalahai, where dumping occurred, villagers reported that their teeth had begun to fall out, their hair turned white at unusually young ages, and they suffered from severe skin and respiratory diseases. Children were born with soft bones, and cancer rates had skyrocketed. Official studies in this village confirmed that there were unusually high rates of cancer, osteoporosis and skin and respiratory diseases. The nearby lake's radiation levels are ten times higher than in the surrounding countryside.[175]

According to a recent MIT study, as the wind industry grows so will the demand for neodymium, which could increase by as much as 700 percent over the next 25 years, and for dysprosium, which could increase by 2,600 percent, The Daily Mail stated that every wind turbine contributes to "a vast man-made lake of poison in northern China."

A huge amount of carbon dioxide is generated in manufacturing, transporting and installing wind turbines. A tremendous amount of concrete goes into the installation of one wind turbine. From that standpoint concrete production, which has increased dramatically since 1900, carries a negative environmental impact. The concrete industry is the third

---

174 http://instituteforenergyresearch.org/analysis/big-winds-dirty-little-secret-rare-earth-minerals/

175 http://www.dailymail.co.uk/home/moslive/article-1350811/In-China-true-cost-Britains-clean-green-wind-power-experiment-Pollution-disastrous-scale.html

ranking producer of anthropogenic [man-made] CO_2 in the world (after transportation and electricity generation). Carbon dioxide is produced at two points during the production of cement. 30% of the carbon dioxide emissions occurs in the burning of carbon fuels, primarily coal, to generate the heat required to drive the cement-making process. 70% of these CO_2 emissions comes from the thermal decomposition of calcium carbonate in the process of producing cement clinker.

To produce just one ton of cement - one component of concrete - 780 kg of CO_2 is emitted. The foundation of just one industrial turbine contains over 500 m^3 of concrete, emitting 624,000 kg of carbon dioxide. This doesn't include the mixing, installation as well as the energy needed to transport the concrete to the job site. This fact reveals the hypocrisy of those who claim that wind turbines are good for the environment. So, considering the number of birds these monstrosities kill and the pollution produced during their manufacture, how could anyone who truly cares about the environment approve of their use?

In our schools children are being hit hard by propaganda put out by multinational corporations with a product to sell. They are asked to name turbines and design logos. They are taken to visit wind farms. The message is clear. Build wind farms - or else the planet will suffer and the polar bears and penguins will die! Promoters of wind energy produce child friendly stories about turbines: Tommy the Turbine, Timmy the Turbine, Lofty, etc. This propaganda is designed to put a positive spin on a controversial industry and keep profits flowing from the next generation.

Yet, there is no hard evidence that building wind farms will do anything to combat "global warming". Costs like extra grid connections, foreign parts and workers, pollution caused by the mining and processing of toxic rare earth minerals vital to their operation are not considered. The decommissioning [scrapping] of wind turbines that have worn out after their 10-15 years of useful life are not included in any CO_2 savings calculations, making emission claims a farce.

Industry and utility company analysts say that by 2020, the United States will need 100,000 megawatts of new electricity. Unreliable wind power simply can't meet these demands. To be competitive with other sources of power, wind farms require huge government subsidies [paid for by taxpayers]. They also need vast stretches of land. For example, according to Rockefeller University professor Jesse Ausubel, meeting New York City's electricity needs alone would require blanketing the entire state of Connecticut with these towering turbines.[176]

Ironically, every wind turbine must be backed up by expensive coal or gas power plants that sit idle when the wind is blowing at just the right speeds. These conventional sources must kick in whenever the wind dies down in order to prevent factories, schools, offices and homes from experiencing blackouts or brownouts. This makes no sense.

The country of Denmark serves as an example. In 1998, the government of Norway commissioned a study of wind power in Denmark and concluded that it has "serious environmental effects, insufficient production, and high production costs."

With a population of 5.3 million, Denmark has over 6,000 turbines that produced electricity equal to 19% of its needs in 2002. Yet no conventional power plant was ever shut down even once during this time. Because wind is intermittent and unreliable, the coal and natural gas plants had to be kept running at close to full capacity to meet electrical demand. They couldn't just be turned on and off as needed. So when the wind is blowing just right for the turbines, the power they generate is usually a surplus and is sold to other countries at an extremely discounted price.

Richard S. Courtney explained,

> "Windfarms for power generation provide intermittent power so they merely displace thermal power stations onto standby mode or to operate at reduced efficiency while the thermal power stations wait for the wind to change. They make no significant reduction to pollution, because thermal power stations must continue to use their fuel and produce their emissions while operating in standby mode with reduced efficiency that can increase their emissions at low output."[177]

---

176 http://nucleargreen.blogspot.com/2008/02/renewable-and-nuclear-heresies.html

177 Richard S Courtney, Oct. 26, 2006, Lecture and White Paper presented to

So the best that the wind industry can ever hope to provide is intermittent power.

In an article "Danish Wind: Too Good To Be True?", published in *The Utilities Journal,* David J. White reported the fact that 84% of western Denmark's wind-generated electricity had been exported at a loss of revenue. The large amount of electricity used by the turbines themselves is typically not accounted for in these figures. Yet ironically, when the wind isn't blowing, Denmark must *import* electricity. In the year 2000 that country imported more electricity than it exported.

Added to the Danish electric bill are the government subsidies that support the private companies building the wind towers. Obviously the government is in bed with these companies. Someone is receiving a big cut of the profits under the table. Danish electricity costs are the highest in Europe. White concluded, "It makes no economic sense to progress an expensive and unpredictable power-generating technology in order to see a parallel carbon dioxide reduction goal when the evidence clearly indicates the objective will not be met."[178]

In Germany utilities are forced to buy "renewable" energy at several times the cost of conventional power. In the United Kingdom, *The Telegraph* reported that, rather than providing cheaper energy, wind power costs the electric companies £50 per megawatt-hour, compared to £15 for conventional power, over three times the cost. The wind industry there is worried that the UK is starting to see that it is only subsidies and government requirements on utilities to buy a certain amount of "green" power to finance the wind turbines and is thus a colossal waste of resources.

The same is true of any electrical grid. Wind turbines respond only to the wind, so their contribution to the supply of electricity on the grid is essentially inconsistent and unreliable. The wind may not be blowing when demand is high. Yet when the demand is low, it could be blowing just right. If there is sufficient demand when the wind is blowing, the need for other power plants to supply electricity is reduced. On the other hand, if the wind speed drops when there is still demand,

---

the Institute of Mining & Mechanical Engineers

178 http://www.windaction.org/posts/33-danish-wind-too-good-to-be-true#.WqBPks9x0xA

conventional power plants must quickly ramp up their operation to cover the power loss.

John Hinderaker describes a recent problem where wind power let people down.

> "Here in Minnesota, we are enduring a brutal stretch of weather. The temperature hasn't gotten above zero in the last three days, with lows approaching -30. And that is in the Twin Cities, in the southern part of the state. Yesterday central Minnesota experienced a natural gas "brownout," as Xcel Energy advised customers to turn thermostats down to 60 degrees and avoid using hot water. Xcel put up some customers in hotels. Why? Because the wind wasn't blowing."[179]

Isaac Orr, another energy expert, commented further about the trouble experienced in Minnesota and other states:

> "Days like today perfectly illustrate why intermittent, unreliable sources of energy like wind and solar would have no place in our energy system if they were not mandated by politicians, showered with federal subsidies, and lining the pockets of regulated utilities that are guaranteed to profit off wind and solar farms whether they are generating electricity, or not."[180]

Professor Randy Simmons disclosed:

> "Over the past 35 years, wind energy - which supplied just 4.4% of U.S. electricity in 2014 - has received $30 billion in federal subsidies and grants. These subsidies prevent people from learning the uncomfortable truth of just how much wind power actually costs. They transfer money paid by taxpayers to wealthy wind farm corporations, many of which are units of foreign companies.

---

179 "Why Green Energy is Futile, in One Lesson," Jan. 31, 2019, www.powerlineblog.com

180 www.stopthesethings.com

> "Many people may be familiar with Warren Buffet's claim that federal policies are the only reason to build wind farms in the US, but few realize how many of the companies that benefit most are foreign. The Investigative Reporting Workshop at American University found that, as of 2010, 84% of total clean-energy grants awarded by the federal government went to foreign-owned wind companies.[181] More generally, the beneficiaries of federal renewable energy policies tend to be large companies, not individual taxpayers or small businesses."

The Institute for Energy Research reported that the actual cost of wind energy is about twice that estimated by the Department of Energy. One reason for this is that the useful life of a wind turbine is only about 10–15 years, not the optimistic 30-year figure provided by the government. Therefore, claims by the wind industry that wind power is becoming cost competitive with fossil fuel-generated power are absolutely false.[182]

If wind power was less expensive than - or even merely cost-competitive with - coal and natural gas energy, these subsidies and mandates would be unnecessary. If and when wind power is about the same price as coal or natural gas power, the wind power industry won't need government to force or induce consumers to buy wind power.[183]

Added to these problems are the adverse health effects of living anywhere near these wind turbines. The incessant grinding sounds they make have been known to cause tinnitus [ringing in the ears], vertigo, panic attacks, migraine headaches, sleep deprivation, and even heart disease, according to groundbreaking research published by Dr. Nina Pierpont. Dr. Pierpont, a leading New York pediatrician, studied the symptoms displayed by people living near wind turbines in the United States, the United Kingdom, Italy, Ireland and Canada. Her findings have led her to confirm what she has identified as a new health risk: wind turbine syndrome (WTS)[184]. This is the

---

181 Randy Simmons, "What's the True Cost of Wind Power?" April 11, 2015, *Newsweek Magazine*

182 http://instituteforenergyresearch.org/press/study-wind-solar-5x-costly-existing-coal-nuclear/

183 .cfact.org/2018/05/26/environment-florida-misleads-voters-on-renewable-energy-costs/?mc_cid=2a94bd2c4c&mc_eid=4e33dab1e3

184 http://www.windturbinesyndrome.com/wind-turbine-syndrome/

disruption or abnormal stimulation of the inner ear's vestibular system caused by turbine infrasound [low-frequency noise], the most distinctive feature of which is a group of symptoms which she calls visceral vibratory vestibular disturbance, or VVVD. They cause problems ranging from internal pulsation, quivering, nervousness, fear, a compulsion to flee, chest tightness and tachycardia [increased heart rate]. Turbine noise can also trigger nightmares and other disorders in children as well as harm cognitive development in the young.[185]

Locating wind farms offshore or in unpopulated areas is not a practical solution. This is because electricity must then be transported long distances to reach the cities where the demand is greatest. The farther energy has to travel from the producer to the user the more energy is dissipated en route. This is an expensive power wasting proposition.

Winds are stronger and more constant in offshore sites than on land installations. However, large ocean waves could dislodge these turbines from their moorings unless very expensive mountings are employed. This type of foundation would destroy the sea bottom at which they were located.[186]

In view of the above, the massive subsidies provided to wind power companies are a tremendous waste of tax dollars. Industrial wind farms are ridiculously expensive and destructive to communities and local environments. Fragile ecosystems are destroyed. Birds and bats are killed. Animal habitats are disrupted, and wildlife is driven from the area. Noise and vibrations cause human health problems, including sleep problems, headaches and nervous disorders. Home values and real estate prices are reduced, with some properties becoming impossible to sell. Energy from wind power is not a viable alternative to traditional forms of energy production. It is therefore neither sustainable nor green.

---

185 Independent.co.uk, Aug 2, 2009)

186 Courtney, op. cit.

14. Energy Cost Calculations

Without gasoline for transportation and electricity to cook our food we'd be back in the era where people worked harder physically and relied on slaves and beasts of burden to get jobs done. The true cost of generating electricity, the life blood of our modern society, comes down to several factors which must all be considered when making this important calculation.

These factors are:

1) the cost to manufacture and install the power generating equipment - such as a traditional coal burning power plant or a solar voltaic farm,

2) the amount of land required,

3) the cost of the resources and energy required to produce it - such as coal or petroleum, photovoltaic cells or wind turbines as well as the environmental cleanup from its production,

4) maintenance of the plant,

5) delivery of the power to the customer through the electrical grid, and

6) the cost of scrapping or recycling the equipment after it has worn out.

Each of these important factors is a part of the equation, and if any of these are left out, we won't know the true cost of the electricity provided.

When researching this question, we put key words into a search engine and are presented with several websites. Since most governments are on the alternative energy bandwagon, those websites ending in .gov will color the picture according to whatever the government's agenda happens to be. If the government is pushing or subsidizing a particular source of electricity, its website will adjust its figures to make the energy source appear as attractive as possible. They will most likely do this by failing to correctly include one or more of the above factors in its cost calculation. The same can be said for each industry, whether it is petroleum [called fossil fuel], coal, wind, solar power, or nuclear energy. We need to bear this in mind when looking at these websites. Hopefully we can find unbiased, independent organizations to provide the answers.

This is not to say that none of these industry websites are totally honest. However, it is human nature to use confirmation bias, or mental filters, to view one's own choices in a more favorable light than the competitor's information. But if a third party website, such as nuclear, for example, is comparing the costs of coal versus wind or solar energy, a more valid comparison than that found at the coal, wind or solar websites will be provided. This is because the nuclear advocates are, in this case, standing on the sidelines evaluating these competitors.

15. The Truth About Solar Energy

Photovoltaic cells, made of semiconducting materials such as silicon, convert sunlight directly into electricity. We use them for all sorts of jobs such as powering watches and calculators and road signs that light up. When photons of sunlight strike these semiconducting materials, they knock electrons loose from their atoms. These electrons are then directed through the semiconducting material to a light or motor. This process of converting light into electrical voltage is known as the photovoltaic (PV) effect.

Under ideal conditions, such as a cloudless sky at noon, if the angle of the solar panel is positioned exactly 90 degrees to the direction of photons of energy coming in from the sun, light falling on one square meter of the panel will produce about 1000 watts, or one kilowatt, of power. This is called Watt Peak.

But often it is rainy or cloudy. Only when the sun is highest in the sky at noon, close to the equator, will you actually have ideal conditions. In a roof installation, for sunlight to fall on the panel at 90º, the angle of the solar panel must face south if you are in the northern hemisphere and be at the same angle as the latitude of the installation. During winter months, when the sun doesn't get as high in the sky, and during morning and evening hours the wattage - and thus the efficiency - drops off. And then, of course, at night the sun isn't shining at all. This is the reason why a typical solar panel has an average efficiency of only about one-sixth of its peak rated power.

The average price per watt for solar panels, which has been declining from previous levels, ranges from $2.67 and $3.43. An average-sized installation in the US costs anywhere from $11,214 to $14,406 after solar tax credits are factored in.[187] A typical video on YouTube by someone who installed solar panels on his roof claims that because his system was rated at 4800 watts, this is the amount of electrical power provided by his system. He will use this figure to compute the time needed to recover the amount of money he paid for the system as only 7.1

187 "How Much Do Solar Panels In The U.S. Cost In 2019?" ,www.energysage.com

years, compared to what he would have been charged by the electric company during that time without his photovoltaic panels.

This is absolutely incorrect. If this person is lucky, the efficiency of his system might be as much as 17% of that provided by peak wattage. In that case he is actually getting about 816 watts, not 4800 watts, out of his system. The time required to recover the cost of this system would thus be about 42 years. Unfortunately, the average life expectancy of such a system before it has to be scrapped is only about 21 years. Therefore, even with the 30% tax subsidy received for buying the system, a person would never recover the investment he made in it. And this doesn't even include the substantial cost of having it scrapped after it wears out.

Therefore, compared with current "fossil fuel" or hydroelectric technologies, solar power is quite expensive. A consumer investing in this technology gets back a return of about 12 cents for every dollar spent. The good news is that the cost per watt is dropping. Between 2010 and 2015 this cost dropped from $7 to about $4/watt.

Large solar projects consume a tremendous amount of land area. To produce enough electricity for a practical large-scale system a project may need several acres of land for each megawatt of power. In all environments—plains, deserts, and mountains alike—large PV projects have negative impacts on important plant and animal habitats and can also pose threats to cultural and archeological sites. Contrary to what we are told, this technology is not environmentally friendly. Large solar projects also require transmission lines, which can present their own environmental challenges.[188]

Another type of commercial solar electricity generating plant is a Concentrated Solar power plant, also known as solar thermal. In this type of arrangement each of an array of thousands of mirrors surrounding a receiving tower is positioned at the precise angles needed to reflect sunlight onto the tower. The Gemasolar power plant in Spain consists of 2,650 such mirrors and occupies 457 acres of land. It produces 19.9 megawatts of electrical power, enough for about 25,000 homes.

188 www.ucsusa.org/clean-energy/renewable-energy/how-solar-panels-work#.WqG9eM9x0xA

Sierra Sun, a plant in the Mojave Desert in California, has three towers and 175,000 mirrors, is rated for 270 MegaWatts (million watts) of power and provides energy for 353,000 homes.

Because the sun's position is constantly changing due to Earth's rotation, each of these mirrors must have its own motor to change its position in order to keep the sunlight focused on the tower. There is either water or molten salt at a temperature of about 500°C. circulating through pipes in the central tower. This molten salt moves through a heat exchanger, where it heats water to steam, powering an electrical generator. The advantage of this type of arrangement is that some of the hot molten salt can be kept in an insulated tank to be used for heating water at night or on cloudy days. This type of energy storage is called a thermal battery ["thermal" meaning "heat"]. This type of operation doesn't need to be backed up with a traditional electrical power plant as do traditional photovoltaic plants to prevent power blackouts.

Some problems with this type of plant are, first, that they are extremely expensive. Second, they take up a tremendous amount of acreage, which impacts the environment. That land could have been used for wildlife habitats or for farming. Maintenance costs to keep such a plant running are high, because each mirror must be at peak reflectivity and exactly the right angle. The best places to locate these plants are deserts, but two problems arise from this. First, water, scarce in a desert, is required to operate such a system. Deserts are often located farther away from population centers than are traditional power plants. Electrical power is lost in transporting it over long distances.

There is also a problem with feeding solar generated energy into the electrical grid. Although peak solar generation occurs in the early afternoon when the sun is highest in the sky, peak demand for electricity doesn't occur until the early evening hours. This mismatch between peak output and peak demand has been dubbed the "duck curve". In the evening, as solar generation decreases due to the sun's position in the sky, system operators must increase output from other energy sources, such as coal, natural gas or nuclear, in order to prevent power outages and brownouts . Therefore, experience has shown that the maximum input to the grid from solar sources can't be any more than about 6%.[189] Exceeding that amount actually has a negative effect on the power grid, which, in that case, could become overloaded.

The cost of these projects doesn't include government subsidies, which are paid by the taxpayer. The Solar Investment Tax Credit (ITC) amounts to a 30% credit on federal income tax for solar projects. Zealous proponents of solar energy keep this part of the cost out of the equation. Taxpayers are not informed that they are actually footing a substantial part of the bill for these projects.

Many states also provide tax credits for large solar installations and, in addition, impose penalties on power companies which fail to include a specified percentage of the power they produce from more expensive wind and solar sources.[190] There is a double whammy here. First an additional burden is put on taxpayers to pay for the subsidies. Second, because wind and solar power systems are so expensive, the power companies have no other option but to pass this increased cost of electricity onto their customers.

One of the worst things about solar voltaic panels is that producing them causes great harm to the environment. According to research by Northwestern University and the United States Department of Energy's Argonne National

189 "The Solar Value Cliff: the Diminishing Value of Solar Power", Aug. 2017, Institute for Energy Research

190 http://dailycaller.com/2016/09/22/the-solar-industry-cashes-in-on-government-subsidies/

Laboratory, as reported in an article appearing in NY Times (2014), "Although China may be a cheaper place than Europe for producing solar panels, the savings come at a higher cost to the environment."

In their analysis, scientists at this lab added together all the costs involved in production: the mining of raw materials, transportation, and the power requirements for production. While China was able to produce these panels more cheaply than in Europe or the US, the greatest cost was harm to the environment. "Sometimes the environmental costs of solar panel production can be lost among the drive to encourage the development of clean energy," declared Huang Xianjin, a Nanjing University professor who studies land use.

Huang and two other professors submitted a letter to the journal, *Nature,* arguing that China needed to take significant action to offset the environmental damage the solar industry is causing there. This would include "improved waste treatment, environmental monitoring and education."

In 2011 a plant manufacturing solar panels in Zhejiang Province halted its operation after residents in the area complained about serious air and water pollution caused by their production. Such reports of environmental pollution are common in China. The Luoyang Zhonggui High-Technology Company produces a highly toxic byproduct of polysilicon used in solar panels -- silicon tetrachloride.

"The land where you dump or bury it will be infertile. No grass or trees will grow in the place. . . It is like dynamite -- it is poisonous, it is polluting. Human beings can never touch it," stated Ren Bingyan, a professor at Hebei Industrial University. The irony here is that solar is touted as green energy.[191]

Meanwhile, because of increasing demand for solar energy, the price of polysilicon soared from $20 to $300 per kilogram in five years. China has been setting new economic records. Dozens of new polysilicon plants are riding the crest of this prosperity. The combined capacity of these new factories is estimated at 80,000 tons, over twice the amount of polysilicon produced in

191 Ariana Eunjung Cha, "Solar Energy Firms Leave Waste Behind in China,"*The Washington Post,* March 9, 2008

the rest of the world.

Unfortunately, due to lax standards, these companies are cutting corners with hazardous waste disposal. Although polysilicon producers in other parts of the world recycle this compound, the Chinese companies prefer to dump it to avoid the high cost of recycling. They've failed to install technology to prevent pollutants from getting into the environment.

Then there is the problem with the disposal of worn out solar panels. In a recent article in Forbes magazine, Michael Shellenberger of Environmental Progress, a nonprofit organization working to promote clean energy, recently expressed his concern about environmental hazards resulting from worn out, discarded solar panels.[192] Typically they contain cadmium, lead and other toxic chemicals which can't be extracted without taking apart the whole panel. In their entirety solar panels must be considered hazardous. Panels left in landfills may break apart and release toxic waste into the ground or bodies of water.

In a 2016 study Electric Power Research Institute concluded, "Solar panel disposal in "regular landfills [is] not recommended in case modules break and toxic materials leach into the soil."

Over a period of time rainwater can wash cadmium out of solar panels and into the environment. This fact was brought up when local people In Virginia rejected a proposal to construct a 6,350 acre solar farm in Spotsylvania County. Sean Fogarty of Concerned Citizens of Fawn Lake stated, "We estimate there are 100,000 pounds of cadmium contained in the 1.8 million panels."[193]

Although there have been some promising advances in solar electricity generation, many more years in research and development are needed to make this type of energy cost competitive and friendly to the environment. To eliminate the need for backup by conventional systems, a technology for storing such energy for use when the sun isn't shining -

192Michael Shellenberger, "If Solar panels Are So Clean, Why Do They Produce So Much Toxic Waste?, *Forbes Magazine,* May 23, 2018

193 https://www.forbes.com/sites/michaelshellenberger/

batteries - must be developed.

Tony Heller of RealClimateScience.com summed up the problem. "The reason winter is cold is because of a lack of solar energy. The sun is low in the sky, days are short, and it is cloudy much of the time. Yet climate alarmists want people to be dependent on solar energy for their survival. They imagine that there is some storage technology which can store huge amounts of energy for long periods of time when the sun isn't shining or the wind isn't blowing."[194]

The Green New Deal and carbon tax schemes anticipate having all of these problems solved immediately and in place within a decade or so. This is an impossible pipe dream entertained only by uninformed politicians and others who don't understand the science or the technology.

194 Heller, op. cit.

16. Grossly Over-Rated Biofuels

There's nothing new about biofuels, or fuels derived from plants such as trees. They've been used since the dawn of history, when people started burning them for cooking and heating their homes. Later, with the mining of coal and discovery of petroleum in the late 1800's, biofuels lost their popularity, especially in developed nations.

The US Energy Information Administration reported that in 2015 a full 43% of alternative fuels were biofuels: burning trees, crops, manure, and trash to produce electricity, heat and fuels for transportation.[195]

One type of biofuel, ethanol [ethyl alcohol] is made from corn or cane sugar. This is the same kind of alcohol found in alcoholic drinks. Other biofuels that have been used are peanut oil, originally used to power the Diesel engine, and hemp, on which Henry Ford's Model-T auto originally ran. During the two world wars in the first half of the 20th century, biofuels were made from several different substances, including potatoes. But when huge supplies of crude oil became available in Texas and Oklahoma, gasoline became a less expensive alternative to biofuels.

In the 1970's the OPEC (Organization of the Petroleum Exporting Countries) nations arbitrarily cut their oil exports to

195 Josh Schlossberg, "[Study] Biofuels Increase Net Carbon Emissions", Aug. 26, 2016, www.thebiomassmonitor.org

raise the price of oil, creating fuel shortages. This caused governments to shift their interest back to biofuels. Part of the reason for this, they said, was that petroleum was a scarce, nonrenewable resource. With more oil fields being discovered every year, we now know that this isn't the case.

The real reason ethanol was added to gasoline in the US, however, was a deal that a group of Democrats led by Congressman Henry Waxman made with the corn lobby. In exchange for its financial support of their political campaigns, they engineered a mandate to blend corn derived ethanol with gasoline as an amendment to the 1990 Clean Air Act.[196]

Advertised as another great source of alternative, clean energy to solve the world's energy problems, ethanol started being added to gasoline in amounts up to 15% without the consent of the consumer. The use of such fuels has been imposed on consumers by this mandate to blend biofuels with gasoline and diesel fuels. It is now difficult in many parts of the United States to find gasoline stations that offer consumers the option of pure gasoline without ethanol. This is true even though gasoline containing ethanol quickly destroys boat and lawn mower engines. Many consumers have had the unfortunate experience of having lawn mowers and rototillers destroyed because they were never warned about the damage the gasoline/ethanol mixture causes in these small engines.

Ethanol-blended gasoline can be devastating to power equipment. It damages rubber and plastic components of fuel systems, especially in older models built of materials that aren't resistant to this type of damage. This includes gaskets, which become brittle, two-cycle carburetors and primer and return lines, which start leaking after a while.

If this type of fuel isn't burned within as short a time as sixty days, it will literally decay. This is because ethanol fuels are hygroscopic. They attract water, and fuel separation can occur and damage the engine's carburetor. Fuel which has decayed in this way contains varnish and sludge deposits.[197] Fuel that is 15% ethanol, called E85, cannot be used in a vehicle unless its engine has been modified.

196 fee.org/articles/the-real-reason-we-have-ethanol-in-our-gas/
197 http://b3cfuel.com/resources/education/the-problem-with-ethanol

This is a worldwide problem. With biofuel mandates now in over 60 nations, the competition between ethanol and food for agricultural land has become a moral issue. Groups like Oxfam and the Environmental Working Group oppose biofuels, because they raise food prices.

Ethanol production has become big business. Millions of acres of virgin prairie and wetlands have been plowed up for the first time ever. The increased use of nitrate fertilizers in growing these crops has caused massive runoffs of nitrates into creeks and rivers, contaminating water supplies in Midwestern states. Cleaning up this water has become very costly.

Now lobbyists for this industry are pushing for legislation which would raise the maximum percentage of ethanol allowed in gasoline from 15% to 30%.[198] This study concluded that 4.2 million acres of additional land is now being used for such farming practices, much of it being land that was formerly native prairie. This is an area that is double the size of Yellowstone National Park.

A gallon of gasoline delivers 116,900 BTU's [British Thermal Units] of energy, whereas a gallon of ethanol delivers only 76,000 BTU's of energy. This works out to about 2/3 the amount of energy from ethanol as gasoline delivers. So a car gets fewer miles per gallon of gasoline/ethanol blend as it does on pure gasoline. Yet, according to David Pimentel in an article published in the Harvard Review, it costs $3.95 to produce one gallon of ethanol from corn. He also reported that "nearly 9 billion gallons of ethanol is produced in the United States, using about 33 percent of US corn grain. These 9 billion gallons of ethanol represent only 1.3 percent of total oil consumption in the United States."[199] Yet at the same time almost eight hundred million people in the world are malnourished.

"The UN's Food and Agriculture Organization (FAO) reported a 45% increase in the world food price index during the 2008 food crisis. Wheat prices increased by 130% relative to 2007 levels. Similarly, soy prices went up by 87%, rice prices by 74%, and maize prices by 31%." Causes of the crisis include the

---

198 Daniel Strohl, "EPA proposes increasing ethanol content in fuel to 30 percent," https://www. hemmings.com/blog, May 17, 2013

199 David Pimentel, http://hir.harvard.edu/article/?a=1911

growing demand for biofuels.[200]

The price of food has also increased dramatically due to unusually cold weather ruining crops.[201] The use of ethanol has had a large part in this and needs to come to an end.

Government subsidies for the corn industry include tax breaks, grants, loans, and loan guarantees.[202] The amount of such subsidies has been estimated at between $1.42 and $1.87 per gallon of ethanol. All the tax credits and other subsidies to those involved in these programs raises everyone's tax bill.

In 2007, as part of the Energy Independence and Security Act (EISA) passed by Congress, the Renewable Fuel Standard version 2 (RFS2) committed the United States to the production of fifteen billion gallons of corn ethanol per year by 2015. The University of Minnesota and the University of Wisconsin, in conjunction with the National Wildlife Federation, conducted a comprehensive analysis of land-use change across the US as a result of the passage of this bill. The leader of this study, Chris Wright, expressed his concern that native grassland was being converted to agricultural land, which has a negative impact on the environment.[203]

The Environmental Protection Agency noted in 2011 that this practice is causing increased soil erosion and loss of wildlife habitat. These are unfortunate and unintended consequences of the use of this type of alternative energy. Yet the EPA extended the year for the mandated fifteen billions of gallons of ethanol to 2017.[204]

"It is now clear that the federal corn ethanol mandate has driven up food prices, strained agricultural markets, increased competition for arable land and promoted conversion of uncultivated land to grow crops. In addition, previous estimates have dramatically underestimated corn ethanol's greenhouse

200 "High Food Price Crisis," www.theoaklandinstitute.org

201 "Other Crop Losses Globally Due to Cold and Mini Ice Age Climate Intensification," May 3, 2017, www.oilseedcrops.org

202 Farzad Thaeripour & Wallace Tyner, "Ethanol Subsidies: Who Gets the Benefits?", April 13, 2007, Perdue Univ. Dept of Agricultural Economics

203 Chris Wright, et. al. "Recent Grassland Losses Are Concentrated Around U.S. Ethanol Refineries, April 2017, Univ. of Minnesota, Duluth

204 http://environmentalresearchweb.org/cws/article/news/68212

gas emissions by failing to account for changes in land use."[205]

These ill advised federal biofuel policies have also adversely affected farmers who raise livestock and related businesses. Consumers are paying for this through increased fuel prices resulting from having to fill up more often because of decreased engine efficiency and miles per gallon.

It is downright immoral to use agricultural land that could be growing food for fuel. The revolution in Egypt called "Arab Spring" started with riots caused by food shortages that were directly caused by this newly instituted practice.[206] But then, believe it or not, "human-caused climate change" was blamed for these rising food prices.

Herr Jean Ziegler commented, "It is a crime against humanity to convert agricultural productive soil into soil which produces foodstuff that will be burned as biofuel."[207]

Biodiesel fuel is a blend of about 80% petroleum diesel and 20% biodiesel, which is produced by treating soybean or another oil with methanol and a catalyst. This industry receives substantial government subsidies from the Biomass Crop Assistance program (BCAP 9010). Producers are reimbursed up to 50% of their cost with payments extending for up to five years for herbaceous feedstock and up to 15 years for woody plants. Section 9003 of the Biorefinery Assistance Program guarantees loans to producers for the construction, development, and retrofitting of commercial-scale biorefineries.[208] Here again, these alternative sources of fuel are viable and competitive only by placing a substantial burden on the taxpayer. Since there is a surplus of petroleum and natural gas, there is no need for ethanol and other biofuels. The government policies responsible for this scam represent the epitome of incompetence, corruption or a combination of the two.

In some states such as Massachusetts forests are clear cut, and the trees are ground up and burned to produce energy. These

205 "Ethanol's Broken Promise: Using Less Corn Ethanol Reduces Greenhouse Gas Emissions", May 29, 2014, www.ewg.org

206 "Did Food Prices Spur the Arab Spring?" https://www.pbs.org/newshour/world/world-july-dec11-food_09-07, Sept. 7, 2011

207 "UN Right to Food Rappeuteur," 2007

208 Alternative Fuels Data Center, U.S. Dept. of Agriculture

forests are replanted with just one or two species of trees, reducing biodiversity from 10-20 species of trees to just one or two species. This practice is harmful to the environment in many ways, disrupting the balance of nature in these habitats. Forests owned by the state of Massachusetts "are being logged at drastic, nearly clear-cut levels and at unsustainable rates."[209]

"Biomass is an appealing 'easy out' for wasteful energy policies and land barons hungry for quick cash. This sweetheart deal for big landholders is being sold as 'carbon neutral', when the facts show this to be a carbon hoax. Taxpayers are lured in by promises of cheap energy and reduced taxes on the myth that this practice is sustainable or healthy for forests, people or the climate."[210]

The combustion of this biomass is more harmful to the environment than the combustion of coal is. Burning wood-based materials releases several toxic organic compounds, including those that produce acid rain. These harmful emissions include carbon monoxide (a deadly gas not to be confused with life-giving carbon dioxide) formaldehyde, benzene and dioxins, all of which are carcinogens. Some of the polycyclic organic hydrocarbons particulates released have been known to cause asthma and other respiratory diseases as well as cancer.

When these biofuels are burned, just as with gasoline and coal, carbon dioxide is released. Biofuel advocates claim that the combustion of these plants is "carbon neutral" due to their absorption of carbon dioxide while they are growing. But it takes 40 to 70 years to replace most of the trees that are being cut down. Meanwhile, the carbon dioxide released by their combustion is already in the atmosphere. Since the IPCC says we have only twelve years to save the planet, this strategy obviously won't help in the least to prevent such a catastrophe. A 2016 study concluded that biofuels actually increase net carbon emissions.[211]

Even worse, when these trees are cut down, the soil microbes located underneath the trees which have been cut release just as much carbon dioxide into the atmosphere as does the

209 R. G. Cachat, "Massachusetts: the Hoax of Biomass and Modern Forestry", Aug. 6, 2016, www.the biomassmonitor.org

210 Ibid.

211 Josh Schlossberg, Ed., thebiomassmonitor.org

combustion of the trees themselves. It takes an estimated 100 years for these soil microbes to recover from this ecological insult. This means that when a tree is cut down and burned, twice as much carbon dioxide is actually released into the atmosphere as burning the equivalent amount of coal.[212]

In order to comply with new regulations calling for the use of "renewables" in the United Kingdom, Drax, its largest electrical generation plant, is now burning wood shipped 3000 miles from forests in North Carolina. Under European Union rules, wood imported from outside the EU is considered to be a renewable, zero-carbon fuel source. In concert with the European leaders, the United States Congress declared nearly all forest biomass to be carbon free. These bureaucrats and politicians need to understand that declaring a substance "carbon free" doesn't make it carbon free.

In the "green" bureaucrats' flawed energy accounting system, this practice is considered superior to the combustion of coal that could have been mined in the UK. The energy required to cut down the forest, convert the wood into pellets and transport the huge quantity of wood needed to run this immense facility have obviously never been considered.[213] Yet this practice has been hailed as a great achievement.

In setting the goal of doubling Europe's 2015 renewable energy

---

212 Ibid.
213 Rupert Darwall, "Tesco and the Great Green Scam," https://blogs.spectator.co.uk/2017/05/tesco-great-green-scam/

levels by 2030, the European Union decreed that cutting down forests to be burned in electric power plants instead of using coal or natural gas qualifies as zero-carbon, renewable energy. They did this even after nearly 800 scientists warned them that this policy would actually increase CO_2 emissions.[214] This Renewable Energy Directive (RED) is now finalized. It is obscene that a huge quantity of wood - perhaps the same amount as for all other uses - will now, due to such brainless policy decisions, be required to meet just a tiny fraction of Europe's energy requirements. How could this practice ever be construed to be helping the environment?

Ironically, to produce the same amount of energy, wood burned in this type of power plant typically emits 1.5 times as much CO_2 as coal and three times as much CO_2 as natural gas. This is because of wood's carbon bonds, water content, and the lower temperatures at which wood burns. After the forest is cut down, the trees will no longer be there to absorb the CO_2 they took in from the atmosphere when they were alive. Whether or not the extra carbon dioxide that will now be in the atmosphere will affect the climate, this regressive action represents a significant flaw in the knowledge and judgment of foolish bureaucrats making such decisions. It's been estimated that replacing coal and natural gas with wood could result in a CO_2 load 2-3 times greater than carbon-based fuels by the year 2050.[215]

The signal the European Union sent to the rest of the world's people with this feebleminded policy decision will have even greater adverse effects on the environment. At the last global climate conference[216] countries with tropical forests, including Indonesia and Brazil, jointly declared their goals "to increase the use of wood...to generate energy as part of efforts to limit climate change."

According to Dr. Patrick Moore, if, through legislation, the combustion of all fossil fuels were to end tomorrow, 99% of all transportation would end, and shortly thereafter "there wouldn't be a tree left on Earth." People would cut them all down and use them to cook and heat their homes.[217]

214 Searchinger, et al., op. cit.
215 Ibid.
216 UNFCCC-COP 23, Bonn 2017
217 Patrick Moore, PhD, Interview with Stefan Molyneux, Sept. 2014

17. Nuclear: Our Energy Future

The combustion of coal releases mercury, a toxic element, and sulfur, which reacts with water in the atmosphere to produce sulfuric acid.

Nuclear energy from uranium carries with it the threat of meltdowns or other accidents. The accidents at Three Mile Island, Chernobyl and Fukushima have demonstrated this all too well. In addition, the disposal of the nuclear waste from these generating plants has never been satisfactorily worked out.

Fortunately there is another type of nuclear energy that is cheaper and safer than conventional methods of nuclear energy production which use uranium. This method of electrical generation begins with thorium, another radioactive element, which is #90 on the Periodic Table of elements. (Uranium is #92.) The way nuclear power works is simple. The heavier elements, including radioactive elements, were produced billions of years ago when huge stars bigger than our sun exploded. The dust from these unbelievably massive explosions called supernovae coalesced into huge clouds of dust, called nebula. Five billion years ago one of these nebulae condensed further, eventually becoming our sun and the planets and moons of our Solar System.

Inside the nucleus, or center of every atom, there is a war going on between two opposite forces: the electromagnetic force and the nuclear strong force. The electromagnetic force is the same force that makes opposite ends of magnets, called poles, stick together. Like poles of a magnet, oppositely charged electrical particles called positive (+) and negative (-) are attracted to each other. Also like magnetic poles, electrical charges that are alike (eg + and + or - and -) repel one another.

The atom's nucleus contains positively charged particles called protons and particles with no charge, called neutrons. The positively charged protons in the nucleus strongly repel one another and would fly apart were it not for another force called the nuclear strong force. This force, which works only at very short distances, holds the protons together with the help of the neutrons. If it were not for the nuclear strong force, atoms wouldn't exist.

Whereas most elements have stable nuclei, and their protons are held together with no problem, the nuclei of very large atoms with either too many protons or the wrong number of neutrons will occasionally lose or gain particles in order to become more stable. This process, called nuclear decay, actually causes atoms of one element to change into those of another element. It is the decay of radioactive elements like uranium that causes most of the heat in the Earth's core and mantle.

In the 1930's it was discovered that when a neutron is shot into the nucleus of certain radioactive atoms such as uranium-235, it splits into two smaller elements, e.g. barium and krypton. In the process two neutrons are released and can smash into other uranium atoms, causing them to also split. In close quarters this can cause a chain reaction which happens lightning fast.

Each time an atomic nucleus is split, a tiny amount of matter is converted into a tremendous amount of energy -- almost a *trillion* times as much energy as the matter from which it was converted. This follows Einstein's equation $E = MC^2$. "E" is the energy produced, "M" is the mass that is converted into energy, and "C" is the speed of light, 300,000 kilometers per second. When you square that large number (300,000 X 300,000), you get an unbelievably large number: almost a trillion. Nuclear energy produces millions of times as much energy per gram as breaking the chemical bonds in the combustion of coal or natural gas.

Just as in coal burning plants, the heat energy produced in nuclear reactions is used to heat water to produce steam. Because steam occupies over a thousand times as much space as liquid water, it exerts a tremendous force that can be used to push a turbine around and around. Because water's boiling point is just 100°C, it has to be kept under tremendous pressure, about seventy times as much as atmospheric pressure. The turbine turns a huge electromagnet. The resulting rapid change in the magnetic field induces [coaxes] electrons to move through a wire to generate electricity. This remarkable process was discovered by Michael Faraday in 1831.

To provide a lifetime's supply of power for the average American, tens of tons of coal must be burned. To provide the same amount of power, just a few kilograms of uranium is needed. However, there is a problem. Over 99% of the uranium found in the Earth's crust is U-238, a form of uranium that can't be used directly, because it can't be split as U-235 can. We say that U-235 is fissile, meaning it can be split by hitting its nucleus with a neutron.

Unfortunately, over 99% of natural uranium is the U-238 isotope. In order to make fission work, some of this U-238 must be removed from the mix to raise the percentage of U-235 from less than 1% to 3%-5%. This process, called enrichment, is complicated and costly. The U-235 isotope thus turns out to be extremely rare and expensive.

In addition to the expense of traditional nuclear fuel is the danger of keeping the water used in these reactors at the pressure required to keep it from turning into steam at 100°C. Just as with a pressure cooker, if this water accidentally escapes as steam, it can no longer do its job of coolant and moderator. In this case, the reaction would be out of control, causing a meltdown of the fuel and the release of radioactive material. This is why nuclear reactors must be housed in huge, expensive containment buildings.

At Chernobyl in the Ukraine there was a huge explosion in 1986 that allowed nuclear radiation to escape from the reactor and affect the surrounding environment. This accident happened during a test in which too many control rods were removed by the plant's operators. The reaction quickly got out of control and exploded. Eight tons of radioactive material were released to the surrounding area, with the city of Belarus receiving the most contamination. Within a few days 30 plant workers died from radiation poisoning.[218] Thousands of workers quickly built a concrete sarcophagus around the reactor, which still contained 200 tons of radioactive material. Initially 115,000 local residents were evacuated. Eventually that number rose to over half a million.

Later, due to an international effort, another huge structure, completed in 2017, was built around the original, hastily constructed sarcophagus. This is expected to contain the radioactive material for 100 years.

The three meltdowns at the Fukushima Daichi nuclear plant in Japan in March of 2011, occurred when one of the largest earthquakes in recorded history caused a huge tsunami to hit the coastal city of Fukushima. The power plant was built at the edge of the ocean so that ocean water could be used to cool the pipes containing hot water that was part of the system.

This tsunami wave penetrated about ten kilometers into the land, killing over twenty thousand people. The water leaped over the seawall there and disabled the plant's backup generators, preventing them from cooling down the reactors.

The explosions occurring in these reactors were not from the meltdowns, which remained within the containment structures. They were explosions of hydrogen that had accumulated inside of the reactors after it separated from oxygen in the extremely hot water molecules. Unbelievably energetic as they were, these explosions caused less damage to the immediate environment than did the explosion at Chernobyl. So the

218 https://www.world-nuclear.org/information-library/safety-and-security/safety-of-plants/chernobyl-accident.aspx

containment structures did their job. However, a great deal of nuclear radiation was released into the Pacific Ocean.

An independent investigation held in 2012 determined that Tokyo Electric Power Company (TEPCO), "had failed to meet basic safety requirements such as risk assessment, preparing for containing collateral damage, and developing evacuation plans."[219]

In a traditional nuclear reactor, called a light water reactor, solid fuel pellets of uranium encased in zirconium are inserted into fuel rods. After about 1% of this nuclear fuel is used up - after being converted to strontium and xenon - the pellets swell, cracking their protective coatings. At this point they are unusable and have to be replaced. Unfortunately, no one has yet figured out how to permanently dispose of the dangerous radioactive waste in this unused fuel. So in the sixty years that nuclear power has been generated this radioactive waste has been sitting around at these nuclear power sites.

Because it isn't fissile, an atom of U-238, which makes up about 96% of this enriched fuel mixture, doesn't split when hit by a neutron. Instead its nucleus absorbs the neutron, changing it into another radioactive element, plutonium-239. Plutonium *is* fissile and is also toxic and very dangerous. This most dangerous of all the elements, plutonium, is the material of which nuclear bombs are made. Plutonium can also be used to power a reactor. So uranium nuclear reactors are a breeding ground for plutonium, which can be used either in nuclear reactors or bombs.

The nuclear meltdowns at Chernobyl and Fukushima made many people distrustful of nuclear power. As a result very few nuclear generating stations have been built in the past thirty years. Many of the older uranium reactors are wearing out, and some are leaking. The good news is that the latest nuclear reactor designs are safer and, in some cases, even able to use some of the waste nuclear fuel. This makes nuclear power more affordable.

The three major nuclear accidents: Three Mile Island (1979), which killed no one, Chernobyl (1986), which killed 30, and

219 https://en.wikipedia.org/wiki/Fukushima_Daiichi_nuclear_disaster

Fukushima (2011), which killed no one, killed a total of thirty people. Considering the amount of electricity generated by nuclear reactors, there have been fewer casualties with nuclear power than with any other source of energy.

A totally different nuclear process which was successfully tested for about five years (from 1965 to 1969) uses thorium, another radioactive element. Because thorium, which is atomic number 90, takes longer to decay than uranium, it isn't nearly as radioactive. The half life of uranium is 4.5 billion years. This means that it would take 4.5 billion years for half of a sample of uranium to decay into another element. The half life of thorium is over 14 billion years. You can hold a lump of thorium in your hand and it won't hurt you.

When hit by a neutron, a nucleus of a thorium atom doesn't split, but just like uranium-238, it absorbs the neutron. This starts a series of nuclear reactions in which thorium is changed into another isotope of uranium, U-233, which ***is*** fissile. When hit by a neutron, U-233 splits into two smaller elements, releasing a tremendous amount of energy in the process. When being split, each U-233 atom releases two neutrons, which, in a chain reaction, smash into other U-233 atoms.

Thorium is four times as abundant as uranium. Even better, nearly all of it can be used without the expensive enrichment process. Thorium has just one isotope, Th-232. 100% of the thorium that is mined - not <1% as with uranium – can be used for fuel. Because thorium is not used as a solid fuel but instead is dissolved in molten salt, 99% of it can be used in the reaction, compared with only 1% of a solid uranium fuel pellet that swells and cracks when the daughter elements are produced. For all of these reasons, thorium can produce electricity about a thousand times more inexpensively than uranium.

Instead of water, the molten salt in which the thorium is dissolved is the coolant in this reaction. Salt has a much higher melting point than water, about 700°C. - 1000ºC., as compared with 0°C for water. So this system doesn't have to be pressurized, as water does in a uranium reactor, making it much safer. As a result, no containment structure is needed for this type of reactor. The power plant can thus be made much smaller and is less expensive to build than a uranium reactor.

In the event of a power failure, the safety system works automatically. Not requiring human intervention, it is foolproof.

Above is the first experimental thorium molten salt reactor, built by NRG, a Dutch nuclear materials firm, in 2017.

An accident such as the one occurring at the Fukushima Daiichi plant - where the backup diesel generators failed - could never happen with a thorium generator. Underneath the system of pipes through which the molten salt circulates is a "plug" of salt that is kept solid by a fan blowing cold air on it. If the power should ever fail or is turned off, the fan will stop running and this plug will melt, allowing the hot molten salt mixture to drain out into larger concrete tanks below. At this point the reaction will cool down and stop, because the atoms aren't close enough together for a chain reaction to occur. So you could never have a meltdown such as those which occurred at Chernobyl, Fukushima and Three Mile Island.

Thorium is so abundant in the Earth's crust that we would never run out of it. A lump of thorium about the size of a tennis ball could provide all of the energy a person needs for a lifetime: for transportation, heating, cooking, light, etc. And this would cost only about $100. So compared with today's cost of energy this amounts to $1 - $2/year for all your energy requirements.

The low cost electricity such reactions would produce could be used to break up water molecules to produce low cost hydrogen gas. We could thus have hydrogen-fueled automobiles. When hydrogen burns, the only exhaust produced is water vapor. The low cost power obtained from thorium reactors could also be used in desalination plants to provide all the fresh water people need. Desalinated water could also transform Earth's deserts into fertile farmland.

Just one mine could provide all the world's energy needs for a year. There are hundreds of such mines worldwide. Because thorium is a waste product of the mining of rare earth metals currently used in electronics and for the magnets used to power wind generators, there are currently piles of it sitting around near these mines.

By now you might be wondering why, if thorium has so many advantages over uranium, it isn't being used instead of uranium to produce electricity. There are two answers to this question. First, the initial research leading to the development of nuclear power occurred during World War II. The leaders of the United States feared that the Nazis in Germany might construct a nuclear weapon. A tremendous amount of money went into the research needed to produce the first nuclear explosion in 1945.

At the conclusion of the war, the focus changed to peacetime uses for nuclear power. Uranium was the fuel of choice for leaders who anticipated the possibility of having to use a bomb in some future war. Some scientists believe that had nuclear technology first been developed during a time of peace, thorium, which can't be used for bombs, would have been used instead.

A thorium reactor which ran for almost five years at Oak Ridge Labs in Tennessee was invented by Dr. Alvin Weinberg, the same man who invented the light water reactor that uses U-235. Believe it or not, when this brilliant man expressed his concern over the safety of the light water reactor, he was fired and the thorium project was canceled.

The nuclear accidents at Chernobyl and Fukushima, along with the realization that wind and solar power are not the answer to the world's energy needs, have inspired a renewed interest in thorium as a nuclear fuel. But there are some technical problems that must be worked out before thorium on a commercial scale becomes a reality.

One problem is that these molten salts in which thorium is dissolved are corrosive at higher temperatures. An alloy containing mostly nickel to house the reactants has been developed. Another glitch is that the element protactinium, one of the intermediate products into which thorium changes before becoming uranium-233, tends to absorb some of the neutrons that are needed to keep the reaction going. Thus, it needs to be removed from the mix at a certain point in the reaction. None of these problems is insurmountable.

Research projects to achieve the goal of a working thorium molten salt reactor are underway in China, India, the United States and the Netherlands, where a working thorium molten salt reactor began experimental operation in 2017.

Another type of nuclear reaction which produces a tremendous amount of energy is fusion. In this type of reaction hydrogen atoms are squeezed together to produce helium.

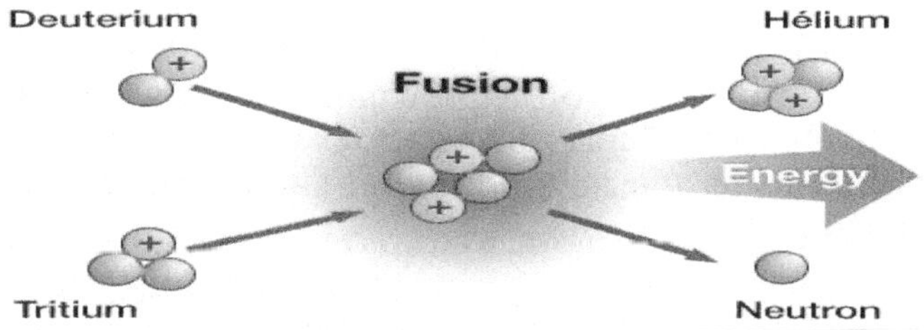

Each hydrogen atom has one proton in its nucleus. These positively charged protons which repel each other, must be put under tremendous pressure at extremely high temperatures to join together to make a helium atom. Because of the heat and pressure required to do this, the reaction is extremely difficult to accomplish. It is impossible to find a reactor shell that won't melt from the extremely high temperatures of the millions of degrees needed in this reaction. So the reactants must be held together by powerful magnetic fields.

It takes a great deal of energy to heat the reactants to the temperatures required to fuse them together. If the energy produced by this type of nuclear reaction isn't greater than the required energy, the process wouldn't be practical for commercial use.

Recently there have been some promising developments in fusion technology. Experimenters at the Massachusetts Institute of Technology (MIT) tested a new type of reactor shaped like a sphere. Rather than trying to join hydrogen atoms together, this process would join hydrogen and boron to produce carbon. These reactants would be heated by lasers.

Meanwhile, in Germany another fusion reactor known as the Wendelstein 7-X Stellerator reactor is being readied for operation. However, this project is reportedly a long way from being able to produce sustainable energy.

At Texas A & M University, a team of researchers experimenting with helium at high temperatures are also finding the task frustrating. When helium heats up to these high temperatures, it begins to bore holes resembling veins in the metal casing.

Despite 60 years and billions of dollars spent on research, fusion on a commercial scale has yet to be achieved and could still be decades away.

Numerous other groups continue to chase the nuclear fusion dream. The largest fusion experiment, located in southern France, is called Iter. Thirty-five nations are participating in the construction of this massive doughnut-shaped fusion generator, called the Iter

Tokamak, which is by far the most complicated machine ever built. This project has been called the most complex scientific endeavor in human history.[220] To hold the extremely hot hydrogen and helium atoms together at 150 million degrees, large superconducting magnets must be used. Strangely, in order for the magnets to be superconductive, their temperature must be held at -269°C. So some parts of the machine are exceptionally cold, while others are exceptionally hot.

220 https://www.theguardian.com/environment/2017/dec/06/iter-nuclear-fusion-project-reaches-key-halfway-milestone

18. Sustainable Hydroelectric and Geothermal Energy

Another type of electricity generation, hydroelectric, uses the force of running water as gravity pulls it down from high to low places to push on the blades of a turbine that turns an electromagnet. The changing magnetic field created by the spinning north and south poles is converted to electrical energy.

Sometimes natural waterfalls, such as Niagara Falls, are used. Most of the time, however, artificial dams are built to trap water at higher altitudes and then allow it to run down a chute, spinning the turbine. Hydroelectric power is produced in countries such as Canada that have both mountains and abundant water.

Hydroelectric power is the most efficient way to generate electricity. Turbines used in modern hydro power plants can convert as much as 90% of the available energy into electricity. In contrast, the best fossil fuel plants are only about 50% efficient. In the US, a kilowatt-hour of electric power can be produced by hydropower for less than one cent. This is the lowest cost of any method of electric power production. However, the initial construction and startup costs of a hydroelectric power plant can exceed a billion dollars, making it difficult for third-world nations to build hydroelectric projects.

One way that hydroelectric power can be used as a storage

battery for solar power is to use solar panels to pump water from a lower level to a dam located at a higher level. Then when the sun isn't shining and you need electric power, you can use the waterfall to fall on the turbine, activating the generator.

Using the force of gravity on moving water, hydropower is also one of the oldest sources of energy on Earth. The ancient Greeks used a form of hydropower, water wheels, to crush wheat into flour.

The first hydroelectric power plant, using technology invented by Nikola Tesla, was built in 1879 at Niagara Falls. This provided electricity for the nearby city of Buffalo, New York.

Hydroelectric power production is growing by leaps and bounds worldwide. One of the largest hydroelectric plants in the world is located on the Yangtze River in China. The dam is 1.4 miles wide and over 600 feet high. About 99% of all the electricity generated in Norway comes from hydropower. In Canada, Venezuela and Brazil that figure is over 50%. Yet only about 3% of all the dams in the United States are utilizing this technology. The other 97% are used for flood control, irrigation and water treatment.

About sixteen percent of the electricity currently being harnessed is produced by hydroelectric generators. That number is expected to increase as new dams are constructed. Seven percent of the electricity generated in the US comes from hydroelectric plants. This is 98% of all the energy from renewable sources in the United States.[221]

In the event of an electricity blackout, hydropower plants would be able to send electricity to grids, quickly shuttling backup power to where it is needed. Some of these facilities can quickly go from zero to maximum electrical output. Worldwide, hydropower saves the burning of 22 billion gallons of oil or 120 million tons of coal annually.[222]

Yet hydroelectric power plants are far from perfect. They can also cause great environmental damage by flooding large land

221 This is provided that petroleum from hydrocarbons is actually from fossil fuels, which is debatable. See Chapter 2.

222 http://powerup.arcadiapower.com/15-facts-about-hydropower

areas on which natural plants and animals exist as well as converting river valleys once used for agriculture into virtual wastelands.[223] The presence of a new dam can even affect the migration patterns of fish.

Improperly constructed hydroelectric dams can be dangerous and life threatening. The worst disaster associated with the production of electrical energy occurred in 1975 when a dam built in China twenty years earlier was unable to accommodate a huge amount of water from a typhoon that had dumped over 40" of rainfall on the area within 24 hours. The sluice gate that had been improperly constructed couldn't take the water fast enough. A huge wave crested over the dam and rushed downstream at 31 mph to create an instant flood. Approximately 26,000 people died from this flood, and another 145,000 died afterward from famine and disease. Almost six million buildings collapsed in this disaster.[224] The risks of coal mining and nuclear plant meltdowns pale by comparison to this disaster.

Geothermal energy is another viable alternative to the combustion of carbon based fuels. Heat deep in the Earth's crust is generated partly by the decay of radioactive minerals deep in the earth. Some parts of the water table are heated by hot rocks close to concentrations of hot melted rock, called magma, that has been pushed up from the earth's mantle. Sometimes this water is even superheated to steam, which can be used to spin turbines connected to electrical generators.

One such operation in the US is a plant north of San Francisco called the Geysers. This plant produces almost one GW (billion watts) of electrical power. Over thirty nations have geothermal electrical power plants. Geothermal heat is also used for directly heating buildings. Iceland, where there are many volcanoes, uses geothermal heat for heating huge greenhouses for growing crops, in addition to electrical power generation.

But there are problems even in extracting heat from the Earth in this manner. For example, a Swiss newspaper article dated December 10, 2009, describing a geothermal project in

223 Krauskoopf and Beiser, *The Physical Universe, 15th Ed.,* McGraw Hill, 2014, p. 117.

224 William Tyrell, "Engineering Disasters: Banqiao Dam Failure," https://www.engineeringclicks.com/banqiao-dam/, July 20, 2016

Switzerland states:

"The authorities in canton Basel City say they will cancel a geothermal energy project, which three years ago caused minor tremors that damaged many buildings. A risk analysis study published on Thursday found that the danger of setting off more earthquakes was too great if drilling at the site resumed."

"The project was put on hold three years ago after thousands of claims for damage were filed with insurers. Total costs for the damage were around SFr9 million ($8.78 million)."[225]

Recently this project has been reconsidered and will probably be completed. Ways will be found to deal with these risks. An existing plant in nearby Riehen will be expanded, and other nearby communities are planning to build similar geothermal projects.[226]

Following a massive eruption of the Kilauea volcano, The *Washington Post* reported in May 12, 2018:

> "Long a concern for residents and the target of lawsuits challenging its placement on an active volcano, the Puna Geothermal Venture (PGV) is a major safety issue in the wake of the eruptions and earthquakes that have shaken the Big Island for days, government officials say." The wisdom of placing a geothermal plant on an active volcano has been seriously questioned for many years by people who understand the possible consequences of such an endeavor."[227]

A safe and totally different way of extracting heat from the ground uses a special type of heat pump called a ground source

---

225 Tyler Durden, "Stunning Images From Space As "Widening & Advancing" Lava Flow Reaches Hawaii Power Plant," https://www.zerohedge.com, May 27, 2018

226 Alexander Richter, "Geothermal energy is seeing strong interest in the three-country region of Basel/ Switzerland, Baden/ Germany and Alsace/ France, with several geothermal heating (and power) projects being kicked off",http://www.thinkgeoenergy.com/low-risk-geothermal-energy-sees revival-around-basel-switzerland/, Feb. 18, 2019

227 Breena Kerr, "Hawaii's Volcano Is Now Threatening a Geothermal Plant Full of Dangerous Chemicals,"*The Washington Post,* May 14, 2018

heat pump. This type of heating/cooling system is also commonly referred to as geothermal, a complete misnomer. A ground source system doesn't have to be located where high underground temperatures exist. It can be located virtually anywhere, taking advantage of the fact that, whether in summer or in winter, the earth is always a constant 55°- 60° F. (8° - 12°C), depending on the part of the world where you happen to be. A network of underground pipes, through which a mixture of water and antifreeze circulates, transfers this constant underground heat to warm an enclosed space in winter, and to cool that same space in the summer. The earth itself is used as a heat source in the winter and as or a heat sink in the summer. A ground source heat pump is simply a heat exchanger, a simple, yet brilliant concept. A system like this can save up to 70% of the energy cost for a home or business.

19. The Motivation Behind the Global Warming Scare

To sum it up, it's all about money and control. H.L. Mencken noted, "The urge to save the planet is almost always a false front for the urge to rule."

In George Orwell's classic, *1984*, a totalitarian government controlled its people with terror and propaganda lies and even by rewriting history. Some of the slogans used by the Party were: "War is peace. Freedom is slavery. Ignorance is strength." The people who had been carefully taught not to question anything went along with this party line. Now in 2019 comes the frightening realization that the theme of this story is coming true. The public is now being told that the frigid winters we've experienced in the past several years means that Earth is warming.

This is a good example of climate change double talk. A recent CNN report claimed that global warming doesn't always mean warming. "The term 'global warming' confuses people, because it triggers thoughts about warmth, and it sort of lends itself to misinterpretation when it also impacts the cold," declared Mike Hulme of the University of Cambridge. The message from this is "Don't trust your senses or your instruments."[228]

When skeptics point to several straight years of severely cold winters, they are reminded that weather isn't climate. "Not only are severe snowstorms possible in a warming climate, they may even be more likely...These storms draw much of their intensity from the extreme contrast between cold air over land and warmer, wetter air from over the ocean," asserted a NOAA spokesperson.[229] So we are being told to suspend scientific observation.

---

228 Eric Worrall, "Damage Control: CNN Explains "Global Warming" Does Not Always Mean Warming,"https://www.sgtreport.com/2019/03/damage-control-cnn-explains-global-warming-does-not-always-mean-warming/, March 3, 2019

229 Laura Stampler, "Here's Why the Crazy Cold Temperatures Prove Global Warming is Real", *Fortune*, Jan. 20, 2010

Likewise, a heat wave that is entirely within natural limits is used for an article warning us that the planet is warming. So whatever happens: whether it's searing hot or frigidly cold, whether there is too much rain or not enough rain, the relentless propaganda to which we've been subjected issues warnings that these weather events are all the result of climate change that ***we*** have caused.

What's frightening about this propaganda is that so many people believe it without question. Those advancing this barrage of lies to which the world has been subjected for nearly forty years know that, as Adolf Hitler stated, "If you tell a big enough lie and tell it frequently enough, it will be believed."[230]

Dr. Patrick Moore summarized the situation. "Fear has been used all through history to gain control of people's minds and wallets and all else, and the 'climate catastrophe' is strictly a fear campaign — well, fear and guilt — you're afraid you're killing your children because you're driving them in your SUV and emitting carbon dioxide into the atmosphere and you feel guilty for doing that. There's no stronger motivation than those two."[231]

In the fable of the boy who cried, "Wolf!" the other sheep herders eventually refused to believe him. We were told in 1988 that within twenty years part of New York city would be under water. In 2005 it was predicted that in ten years [2015] gasoline would cost over $9 a gallon and milk as much as $12 per gallon. This also never happened.

But the purveyors of propaganda are banking on the unfortunate tendency of people to forget past predictions that didn't materialize, as described in Chapter 2 of this book. The corrupt media seem to be able to get away with pumping out these false stories nonstop. Now, after countless thousands of more cleverly orchestrated false alerts and tipping points, it seems that some people have more or less *accepted* the concept that climate change is "here" and is "even worse than predicted".

---

230 https://www.brainyquote.com/quotes/adolf_hitler_385028

231 Robert Craychik, "Greenpeace Founder: Global Warming Hoax Pushed by Corrupt Scientists 'Hooked on Government Grants'", *Breitbart News,* March 7, 2019

For example, we read here: "As climate change becomes more severe than originally projected, estimated adaptation finance needs for developing countries are doubling or tripling every few years."[232] (emphasis added)

The term "Green New Deal" was actually coined in 2007 by Thomas L. Friedman, a *New York Times* columnist.

When push comes to shove, since sane people would never volunteer to give up their precious liberty and the wealth that has been achieved through the free enterprise system, only a socialist (or communist) totalitarian government backed up with big guns will ever be able to attempt such a feat.

Don't think for one minute that this couldn't happen. It has been said that those who haven't learned from the past are doomed to repeat it. We only have to remember how the Nazi government in Germany controlled its people with similar lies that led to war and the deaths of millions. The same tactics were used in power grabs by twentieth-century communist governments, which were ultimately responsible for the deaths of 250 million people.[233]

In order to persuade people to do things that are repugnant to them - like accepting a worse standard of living and giving up their freedom - a lot of fear and guilt must be heaped on them. Most people recoil at the thought of war. So throughout history corrupt leaders have softened up their resolve to avoid war by using corrupt, state-run media to manufacture a barrage of false reports about the "enemy". In this case the enemy is "climate change," and we are told that we're in a war.

Just one example of this is the Vietnam war. The truth about how the American people were duped into accepting this war didn't surface for fifty years when documents became unclassified. In a solemn address to the American people in 1964, President Lyndon Johnson reported that US warships had been attacked by North Vietnamese PT boats. This turned out to be a total lie. That staged event that never took place greatly

232 Lisa Dougherty-Choux, "The Costs of Climate Adaptation, Explained in 4 Infographics," https://www.wri.org/blog/2015/04/costs-climate-adaptation-explained-4-infographics, April 23, 2015

233 Christopher Monckton, speech to

expanded US involvement in Vietnam.[234] Over 58,000 young US soldiers and two million Vietnamese were killed in this needless tragedy that made many people rich.

How many people will believe this? Some people still can't wrap their minds around the idea that their government may be lying to them. Cognitive dissonance is the psychological term for such thinking. They prefer to believe that anyone who entertains such notions is a "conspiracy theorist". However, the lie described above is not a conspiracy theory but a conspiracy ***fact***.

As Mark Twain said, "It's easier to fool people than to convince them that they have been fooled."

The founding fathers of the United States set up a republic - a representative form of government - of the people, by the people and for the people. Its power was purposely limited to prevent any one person or group from getting too much power.

Now there are forces within the Democrat Party (and even some in the Republican Party) who are threatening to do away with the Constitution of the United States. The US Constitution, though having been amended many times, is still one of the most brilliant documents ever devised. It launched an experiment that was purposely designed to keep government from getting out of control. This latest threat to the world as well as to the United States is based on the false premise that life threatening global warming is a result of mankind's activities.

234 Steve Watson, Gulf of Tonkin Was a Staged False Flag Event," http://themillenniumreport.com/ 2014/08/senators-knew-gulf-of-tonkin-ws-staged-event-docs-show/, Aug. 10, 2014

20. Nonsolutions to a Nonproblem

The Green New Deal proponents who say they want to save the planet evidently believe they can do this by replacing relatively inexpensive and dependable methods of energy production with three unsustainable types of energy generation: wind power, solar power and biofuels, (see chapters 13, 15 and 16). Nuclear energy and hydroelectricity, both of which are viable alternatives that don't emit carbon dioxide, seem to be off of their radar. The fact that these politicians are offering an unworkable nonsolution to a nonproblem is a testimony to their inexperience and ignorance of what they are talking about.

The Green New Deal plan calls for "dramatically expanding and upgrading renewable power sources" to achieve their climate change goals. That would mean importing huge quantities of not just neodymium, but also other critical minerals such as praseodymium, dysprosium, terbium, indium and silver, to meet the production demands of wind turbines, solar panels and electric cars. China would be a big winner here, since it is the main source of these rare earth metals.

Electric cars also use batteries, which require lithium and cobalt, about 40 percent of which is refined in China. "China already dominates global battery manufacturing and is on track to supply nearly two-thirds of all production by 2020," Manhattan Institute senior fellow Mark Mills wrote in a recent paper.[235]

When the wind isn't blowing at the correct speed and when the sun isn't shining [at night and on cloudy days], people are totally dependent for their survival on carbon based fuels, and nuclear energy and hydroelectricity. That's the reality of the situation. Even Bill Gates recognizes this. He stated, "Solar is only during the day, solar only works best in places where it's warm. There's no battery technology that's even close to allowing us to take all of our energy from renewables and be

235 Michael Bastasch, "Ocasio-Cortez's Green New Deal Could Be a Huge Boon For China. Here's How," https://dailycaller.com/2019/04/05/green-new-deal-china-energy/,April 5, 2019

able to use battery storage in order to deal not only with the 24-hour cycle but also with long periods of time where it's cloudy and you don't have sun or you don't have wind."[236]

Supposing, for the sake of argument, that human generated CO_2, a small fraction of that which enters the atmosphere naturally every day, would produce a warming of an additional 1.2°C. Whether or not this would be a hardship or a boon to humans and other organisms is a matter of opinion. We know from history that during the Roman and the Medieval warm periods civilization flourished and people prospered.

As stated earlier in this book, only about 4% of all the carbon dioxide emissions are produced by human activities. This is a fact. Add to this the fact that the *principal* greenhouse gas isn't carbon dioxide at all but water vapor, which has about 70 times the warming capacity of carbon dioxide. The global warming alarmist establishment has never produced a single study based on real observational science that ties the warming experienced in the last two decades of the 20th century to CO_2 emissions. As stated before, the earth actually *cooled* between 1940 and 1978 while at the same time atmospheric CO_2 was rising.

The precautionary principle, as defined by the New World Encyclopedia, is the idea that we should take action to correct a practice that could harm people or the environment *even when that potential harm hasn't been conclusively proven.* In this case the practice is the combustion of carbon-based "fossil fuels". The precautionary principle places the burden of proof onto those saying the practice isn't harmful. This is "a major principle of international environmental law and is extended to other areas and jurisdictions as well."[237]

This requires policy makers to make discretionary decisions to avoid potential harm until scientific findings prove that that no harm will result. Under strict enforcement of this principle, however, the cell phone wouldn't have been permitted until it was proven not to be a carcinogen. But the burden of proof should be on those who are asking others to spend trillions of dollars to mitigate what may not be a problem at all. Not only do these people lack proof, but they have been caught lying with data (chapters 7 and 8) and concocting many false claims

---

236 Heller, op. cit.
237 http://www.newworldencyclopedia.org/entry/Precautionary_principle

(chapter 5).

The case of the banning of the pesticide DDT is an example of how the precautionary principle has been misused. In 1962 the bestselling book entitled *Silent Spring,* by Rachel Carson, vilified DDT on the grounds that it was harming some species of birds.

A 7-month study in which 125 expert witnesses provided testimony concluded that DDT wasn't harmful. In 1972 EPA director William Ruckelshaus decided to ban it in the US anyway, even though he hadn't attended the hearings or even read the report of the investigation. US foreign aid was withheld from third-world nations which hadn't banned it.[238] At the time this was happening malaria, which was controlled by DDT, had almost been eradicated. Due to this ban, however, there was a resurgence of this deadly disease. The sad result is that now millions of people, mostly children, die from malaria every year.

Collectivism is a political philosophy that puts the welfare of the group before that of the individual. Collectivism is the opposite of individualism, which is the bulwark of the free enterprise system. Although decisions in such a society would theoretically benefit the people, this has never worked in previous experiments, because it goes against human nature. Socialism and communism are two forms of collectivism. They require a totalitarian dictatorship, as seen in the Soviet Union. A new form of collectivism is technocracy, or the dictatorship of technically oriented bureaucrats, who think they know better than other people how to run their lives and are prepared to tell them how to do so.

The Green New Deal proponents cleverly use the emotions of fear and guilt to persuade Americans to put the welfare of society above their own. Using the climate-change nonproblem as an excuse to impose collectivism on the American people has been called "a Trojan horse for a bigger nationalization of the economy than seen under President Franklin D. Roosevelt."[239] The plan's sponsors themselves define their goal as the "massive transformation of our society." Anyone who is sane

---

238 Zubrin, op. cit., p. 102

239 Ryan Bourne, "Alexandria Ocasio-Cortez's Green New Deal Is a Radical Front for Nationalizing Our Economy," *USA Today,* Feb. 11, 2019

would accept such a foolhardy plan only if he or she were convinced that we are in a war for survival. Yet there is overwhelming evidence that the tiny amount of warming experienced in the final 20 years of the 20th century is not a threat at all. This happens to be the same amount of warming as that which occurred between 1910 and 1940 when the level of carbon dioxide was lower.

Statistics show that the increase in atmospheric CO_2 hasn't produced the disastrous consequences the warming alarmists say it has. The amount of warming experienced since the beginning of the Industrial Revolution falls entirely within natural limits and is even less significant than the warming occurring in the Medieval and Roman warm periods. Even more significant: there is no scientific proof that the slight warming of the last century has been caused by mankind.

Because the Green New Deal fails to address strategies, trade-offs or costs involved in implementing it, it is completely impractical. The proposed solutions are so irresponsible that we have to wonder whether its proponents have ever had any training in science or any personal life experience handling money, such as paying bills or working within a budget. Even if informed of the cost, many people would recognize that past predictions made by politicians were usually way off the mark to the extent that the project ended up costing many times the original estimate.

We only need to revisit the ineptitude of the Obama administration in wasting over a half billion dollars of taxpayers' money to provide a sweetheart deal to Solyndra, a solar voltaic panel manufacturer. The unelected bureaucrats who researched that company and made the decision obviously had no business sense or knew what to look out for. When it became obvious that this company was going bankrupt, the Obama administration persuaded Solyndra's leadership to delay laying off workers until after the 2010 election.

Politicians on the state level are similarly clueless. In 2012 the tiny state of Delaware awarded over $25 million in grants to Bloom Energy, a fuel cell manufacturer. In addition, this company paid a rent to Delaware of only $1/year for a huge facility that the state had built for them. Delaware even subsidized worker's salaries to the tune of $6,250/year for each full-time worker (up to $36 million per year.) This should never have been necessary in the first place. If these companies aren't competitive, why should they be given these special advantages? 39% of Delaware's renewable energy budget was even allocated to Bloom Energy to producing energy using natural gas. This meant that fossil fuels had to be designated as "alternative energy". Redefining terms is a complete lie and a sure sign of corruption.

Because of Bloom's inefficient energy production, Delaware residents woke up to discover that their electric bills had been substantially increased.

Bloom Energy had sold itself as a "green" energy company that would produce no hazardous waste and even declared this on its permit request. So this company had been exempted from hazardous waste rules by the state of Delaware. But Bloom Energy's production of fuel cells produced arsenic, benzene, chromium, lead, carbonyl sulfide, hydrogen sulfide, methanethiel and volatile organic compounds. These hazardous chemicals had to be collected in canisters and shipped to New Mexico, which they did illegally.

Bloom Energy also violated a coastal zone rule, because of its proximity to Delaware Bay. So Delaware fixed that problem by rescinding the Coastal Zone Act.[240] It's amazing what one can accomplish with the stroke of a pen. When the governor was asked why he had done this, his answer was "to fulfill the requirements to provide 25% of Delaware's energy from renewable sources." How many other states have similar goals for "renewable" energy?

The truth is that, rather than being green, efficient and affordable, the fuel cells made by this company emitted more pollution, and produced energy which was three times as expensive, as a state-of-the-art natural gas, combined cycle plant. This company actually was emitting more CO_2 to produce the same amount of energy as a traditional plant.[241]

These are just two examples of the bumbling ineptitude and outright corruption that states and the federal government have

240 David Legates, PhD presentation at the 36th Annual Meeting of Doctors for Disaster Preparedness, August 26, 2018, Las Vegas, YouTube

241 Ibid.

demonstrated in handling other people's money. It's all about the money. There are winners, and there are losers. The losers are the public. If this affair had been handled by private industry without government subsidies and other interference, it would have had a much different outcome.

The proponents of such renewable energy schemes take advantage of the fact that most Americans care deeply about the environment. Most people have never heard the other side of the story, which has been blocked from discussion. Saying that "The science is settled," precludes any opportunity for public discussion or debate. It's understandable that even intelligent, educated people could be taken in by incessant propaganda that the warming is caused by human activity and has "social costs". By telling people without furnishing *scientific proof* that the window of opportunity to take action is very small - only twelve years to "thermogeddon" - the IPCC is guilty of treachery of the highest order.

The Green New Deal isn't likely to be accepted by the American people in its present form. Those who made up this wish list of benefits we're promised if we accept the sacrifice it entails - basically giving up our freedom - undoubtedly realized this from Day 1. They threw it out as a *starting point for discussion* to get people used to the idea that, even if not ***this*** plan, ***some kind*** of drastic action, such as a tax on "carbon" (really carbon dioxide) must be implemented. This is a classic strategy for which the Democrat party is famous. Ask for something impossible, and then get people to move in that direction by compromising. A "carbon" tax somehow doesn't seem so bad when compared with the Green New Deal.

With no scientific proof for this hypothesis - which, as we have seen, is impossible to disprove - over a hundred *billion* dollars has already been spent just on climate research alone. Add to that a similar amount spent on renewables research.

The FAQ section of the Green New Deal resolution states that this program would mobilize American society at a scale not seen since World War II. At that time President Franklin Roosevelt set a precedent, ordering the auto industry to replace the production of cars and trucks with that of aircraft engines, tanks, etc. There was a great deal of propaganda used in this effort. The federal government took advantage of an opportunity to grab control, and industry capitulated to its pressure. Many new federal agencies were created, including the War Production Board, which grew to 20,000 bureaucrats. Government agencies were given the power to requisition any property required for the war effort. They also controlled the supply of raw materials. The Green New Deal planners have indicated that they will employ a similar strategy to achieve at least the same level of control and enforcement capability.

The difference is that the United States under Roosevelt was fighting a real enemy.

During his February 2019 press conference, Sen. Markey exclaimed, "From the air we breathe to the jobs that employ us to the neighborhoods we live in, to the economy we operate within, climate change defines our existence." All of these oppressive controls would be carried out in the name of human-caused "climate change".

Some Republicans also - not to be outdone by the Democrats - have introduced their own ideas for using this issue to take advantage of a good "crisis" to suck more tax money out of Americans' pockets. To assess the wisdom of such a move we need only to examine the consequences of such tax plans in Australia and France. In 2012 a $25/ton carbon dioxide tax was introduced in Australia. This ten-year plan would have cost $162 billion. The goal was a modest 5% reduction in carbon dioxide emissions. This caused such extreme economic deprivation that the number of bankruptcies skyrocketed. Because Australia's CO_2 emissions amount to only 1.2% of mankind's total CO_2 emissions, this reduction would have made a difference in the earth's average temperature of only 0.00005°C after ten years.[242]

If we extend those figures out to worldwide proportions, we find that the cost of reducing the earth's average temperature by even one degree Celsius would be $3.2 quadrillion ($3,200 trillion), an astronomical figure. This is assuming that the IPCC's estimates are correct. Reducing the temperature just 1/7 of a degree would thus cost $450 trillion. This cost, spread out equally among the Earth's seven billion human residents, would be $65,000 for each man, woman and child. And that's

242 Address by Lord Christopher Monckton, "Carbon Tax - Lord Monckton - Part 2," Jan. 5, 2012, YouTube video

for just 1/7 of a degree, hardly noticeable. Would the sacrifice really be it worth it?

The alarmists tell us that it is our duty to make these sacrifices for future generations: our children and grandchildren. But if we bankrupt ourselves and wreck our economy, what sort of legacy are we leaving them? It would make far more economic sense to simply adjust to any slight warming that might occur if indeed we are influencing the climate at all.

Dr. Richard Lindzen noted that we don't even know what the Earth's ideal temperature should be. Earth didn't come with an instruction manual to let us know what that is. We do know that during the Medieval Warm Period, which was about 2°C warmer than now, agriculture flourished in Europe and Greenland was green. People were doing much better at that period in history than during the Little Ice Age.

We and our elected representatives need to exercise extreme care not to get suckered in by any kind of cap and trade legislation or by a "carbon" tax, which some tax experts believe would devastate our economy. This extremely regressive tax would place a tremendous burden on the poor, because energy constitutes a much larger percentage of the budgets of the poor than of the rich.

Several proposals for such a tax have already been suggested. Florida congressman Francis Rooney sponsored the Energy Innovation and Carbon Dividend Act, a trillion dollar tax and spend welfare scheme. It would tax carbon dioxide emissions by $15/ton the first year and then increase $10 per ton each year after that. These costs would, of course, be passed along to the consumer. Rooney claimed that because some of that money would be returned to the public in the form of checks, the bill would be "revenue neutral".

Even if, in such a plan, some of this money should be returned to the consumer, it would nevertheless significantly grow the size of government, which is already much too large. Thousands of bureaucrats would have to be hired to administer the tax. This would also penalize families with more than two children. A family of five would receive the same amount as a family of four.

Any tax on carbon dioxide would cause significant inflation, raising the price of all goods the production of which generate CO_2, including agriculture and concrete. Not only would the cost of food skyrocket but the cost of transporting it to grocery stores would also greatly increase.

It is hoped that, after reading the evidence that global warming alarmism isn't based on observation and measurement but on pseudoscientific models, you will want to do your own research on this subject to verify the truth of what's been presented in this book. In another generation or so people will probably look back with wonderment on this strange period of time.

There's nothing wrong with changing your mind about an issue when new data comes to light. Some scientists have expressed their concern that when the truth about this deception does come out in the open, many people who once believed in the integrity of science will no longer trust science for answers. That would be truly unfortunate.

21. The Brainwashing of the Young

For approximately forty years now the AGW establishment has been indoctrinating the public worldwide, and their media accomplices have been only too willing to cooperate. But nowhere is this more evident than in our public schools. Just as the Nazi's brainwashed the German children, the global warming movement, which hijacked environmentalism, has unfairly used their captive audiences in schools to influence their young minds.

On Friday, March 15, 2019, school children all over the world took the day off from school to participate in a demonstration to save the world from climate change. The headmasters of schools in Britain suspended schools even though their students, were supposedly "on strike". It was obvious that this affair, arranged to garner more media attention, was organized by the adults.

Al Gore expressed his highest praise for this action, saying, "Today, hundreds of thousands of young people in over 120 countries stood up to fight for their future. Going on strike from school, they spoke with one voice and one message for the adults of the world: Act on climate. Now...I've been in this fight for many, many years and I don't have words for the pride and joy I feel witnessing these brave young people speaking out today. If you were a young activist striking today, thank you."[243]

This sort of manipulation of young children's minds takes advantage of the fact that children are impressionable and often lack the education and experience to discern truth from fiction. The strategy of the organizers is reminiscent of that used by the Hitler Youth leaders in Nazi Germany.

One of the worst nightmarish examples of this type of manipulation is David Suzuki's statement:

---

243 Mark Morano, "Al Gore: Kids climate strike from school was 'extraordinary' – 'I don't have words for the pride and joy I feel,'" Al Gore: Kids climate strike from school was 'extraordinary' – 'I don't have words for the pride and joy I feel'" March 15, 2019

> "The North Pole, once a wintery wonderland, is no longer safe for Santa's Workshop. Climate change is melting the snow and ice, and the rising water is getting too close for comfort. Santa must relocate — fast — to make sure that all the nice boys and girls still have a Happy Holiday."[244]

We have a great responsibility to teach children as well as adults how to recognize the signs that they are being exploited.

Propaganda is a persuasive and widespread message designed to advance the interests and agenda of a particular group. What sets it apart from other mass communication techniques is that it bypasses logic by using faulty reasoning - logical fallacies - and emotional appeal.

Some logical fallacies constantly used in manmade global warming propaganda are:

1. ad hominem (against the man) attacks - which either directly call names, such as "climate denier" or indirectly disparage those who disagree by assaulting their credentials or character.

2. "straw man" - distort your opposition's argument and then knock it down.

3. "band wagon" - This technique states that there is a consensus of the majority in favor of the agenda, and that anyone who disagrees is on the "fringe", and that there is something wrong with the person who disagrees. This works, because of the basic human need to be on the "winning team".

4. "appeal to authority" - We are told that we need to "trust the experts" without being given any direct information about the truth of the matter. The telltale sign is the lack of information. We are never told that there are other experts who disagree with the "experts".

5. "Moving the goalpost" When Mother Nature didn't cooperate with the global warming models and Earth stopped warming after 1998, the global warming alarmists changed their slogan

---

244 https://junkscience.com/2011/11/suzuki-where-will-santa-live/

from "global warming" to "climate change" and even "climate disruption" to incorrectly imply that this was causing storms, droughts, etc.

6. "An appeal to sympathy" is made by blaming those who disagree for the plight of the polar bear (see chapter 5) and the victims of storms, droughts and floods.

Dr. Tim Ball has identified some telltale signs that indicate that something we read and hear is propaganda.

1. Weather events that are normal are presented as abnormal. This is because the climate change we've experienced in the last century is all within the parameters of natural climate change that is to be expected. The history of such events plainly shows that, despite what the alarmists say, there has been no increase in droughts, floods, heat waves, wildfires, hurricanes or tornadoes.

An example of this type of reporting is this headline:"The 2017 Hurricane Season Really Is More Intense Than Normal." [245] The key phrase here is "more intense than normal".

2. Definitions are changed. Doing this is a logical fallacy known as equivocation. In 2004 climate alarmists, realizing that the warming had stopped, changed their alarm of "global warming" to "climate change".

"This gave them a greater range and made the problem more threatening. Now you see headlines identifying that it was, variably the warmest, coldest, wettest, driest, calmest, and windiest."[246]

3. They play on our emotions - such as fear, guilt and pity - rather than on reason. They imply that we should feel guilty for what we are doing to our children and grandchildren. We should take pity on the poor polar bears.

245 Maggie Astor, "The 2017 Hurricane Season Really Is More Intense Than Normal." *The New York Times*, Sept. 19, 2017

246 Dr. Tim Ball, "Seven Ways To Spot Climate Change Propaganda," http://www.technocracy.news/dr-tim-ball-seven-ways-to-spot-climate-change-propaganda/, Feb. 12, 2019

A good example of this type of propaganda is the following editorial[247] by Mark Hertsgaard:

> "Climate Denialism Is Literally Killing Us. The victims of Hurricane Harvey have a murderer—and it's not the storm -
>
> "The horrors hurled at Houston... in late August were heartbreaking—but also infuriating. How many times must we see this disaster movie—titled Hurricane Harvey in 2017, Hurricane Sandy in 2012, and Hurricane Katrina in 2005, along with many lesser-known foreign releases—before we intervene and change the ending? And how long before we hold the ultimate authors of such climate catastrophes accountable for the miseries they inflict?
>
> "The tragedy of Harvey starts with the suffering of innocents like Jordyn Grace, the 3-year-old who survived the flood by clinging to the body of her drowned mother, who had prayed with her last breaths. At least 60 people died in Texas because of the storm, over 1 million people were displaced, and who knows how many survived but lost everything?
>
> "What makes this so infuriating is that it shouldn't be happening...And how long before we hold the ultimate authors of such climate catastrophes accountable for the miseries they inflict?"

The rant goes on and on. To whip people into an emotional frenzy, he blames this storm on "climate denialism", which itself is a misnomer. No skeptical scientist has ever denied that there is a climate or that the climate is changing. Hertsgaard also makes the totally irrational and incorrect assumption that such hurricanes did not exist before man started putting carbon dioxide into the atmosphere. Nothing could be further from the truth, detailed in chapter 5 of this book, that eight of the ten worst hurricanes in US history occurred in the late 1800's and early 1900's when the level of carbon dioxide was much lower than today.

---

247 Mark Hertsgaard, "Climate Denialism is Literally Killing Us," *The Nation* thenation.com, September 25-October 2, 2017, Issue

A category 4 storm, Harvey was the country's first major — category-3 or higher - hurricane to make landfall on the US mainland since Hurricane Wilma hit Florida in 2005. This ended a twelve-year period, the longest in US history, with no major hurricanes. Evidently Hertsgaard either didn't do his homework or tried to push lies onto the American people in a deliberate attempt to instigate fear and misdirected blame.

To add to the emotional and illogical tone of the conversation, propagandists do outlandish things like giving Mother Earth legal status with the right to sue corporations and "climate deniers".

4. They take advantage of the public's lack of science knowledge. Climate is a very complicated subject. Even some scientists who specialize in other disciplines, such as biology, may lack the confidence needed to make an informed decision. So they trust the "experts". This enables the climate alarmists to make up fake terms such as "polar vortex" and unscientific statements, such as "The science is settled." Real science is never settled.

5. They keep ramping up the threat to make it more and more ominous. The latest IPCC statement now warns that we have only twelve more years to do something about climate change. This is just another of the many tipping points that never came to fruition. (See chapter 2 to learn how many past predictions have materialized. However, we can teach young people how to spot propaganda by allowing them to analyze articles about the subject and spot evidence of these tactics.

Dr. Patrick Moore calls this shameful exploitation of children to advance one's agenda "child abuse". He recently told the climate alarmists in a tweet,"You don't seem to get it, that this is a fascist tactic. You are using innocent children for your own ends. And this is child abuse of a psychological nature. It's normally called brainwashing, and you are guilty of participating in it. Mao would be proud...You are clearly projecting your own psychosis on the children. Self-loathing is a heavy burden. Why don't you express it through adults rather than abusing innocents?"[248]

248 Dr. Patrick Moore, in a March 15, 2019, tweet.

22. A Sustainable Energy Plan

As stated before, the entire "climate change" scenario that was cast upon the world in the 1960's – first as a global cooling crisis and then as a global warming crisis - really takes advantage of the world's need for energy, without which very little can be accomplished. In the not-too-distant past energy was provided mostly by animals and by human slaves as well as by burning wood. Wood, as well as coal, was used in the first steam engines, leading to the Industrial Revolution. Petroleum was discovered in 1859.

As the 21st century progresses, the world's people will need more and more energy. The world currently relies heavily on carbon-based fuels, especially coal, which now provides about 50% of the world's electrical energy. It can't be denied that there are many health hazards associated with coal mining. However, coal-fired plants, discussed in chapter 12, are now much cleaner and more efficient than ever before.

It would be impractical and inhumane to forbid this inexpensive and dependable source of energy that people desperately need ***now*** for heat, transportation, agriculture and food storage. For those in power to deny the world's people access to carbon based fuels without offering them a *viable and sustainable* alternative, such as nuclear energy, would wreck the entire world economy. (But then this is what Maurice Strong said he wanted to do.) Withholding these energy sources from people would sentence millions to death from starvation, freezing and disease.

Many people currently being blocked from access to affordable electricity are needlessly trapped in poverty. Millions of people in India, the Philippines and other developing countries spend several hours every day searching for fuel just for cooking dinner or lighting. This is time that could be used much more productively. Many of these people are women and children, who must walk long distances carrying home water, scraps of wood or animal dung. These women and children could instead be learning skills that would help raise their standard of living.

Poverty caused by the absence of benefits electricity provides prevents people from coming together to produce goods in small businesses, such as sewing factories that could produce useful garments. If there's no reliable source of reasonably priced electricity to supply the sewing machines, this would be impossible.

Many of these people have access to farm land. But, with no refrigeration to store food, much of the harvest can't be preserved, and people go hungry. Harvests are often sporadic. The inability to preserve food from a good harvest prevents people from saving it for the lean times. These people are even being denied the availability of electric pumps for fresh water and irrigation.

Although the availability of power wouldn't, in itself, prevent poverty or end the exploitation of women and children for manual labor, it would help people go a long way towards getting their basic needs met. Being able to cook a meal by turning on a hot plate or to read at night by switching on an electric light would be a starting point to improve the quality of life for the world's poorest people.

As Christopher Monckton so aptly stated, "Efforts to make a nonexistent problem go away would hurt the poor the hardest."[249]

We need to provide the people in third-world nations with the same opportunities that helped the richer nations to get where they are. Energy provided by fossil fuels can be gradually replaced with safe nuclear reactors.

Back in the 1970's the public was told that in another twenty years the world would run out of petroleum. Since that time, however, more and more oil deposits have been discovered. There are now about 65,000 known oil fields, many of which are gigantic. Over 900 of these contain at least five hundred million barrels (about 79,000,000 cubic meters) each. The oil companies don't want the public to know just how much petroleum there is and how cheaply it can be pumped out of the ground. Nevertheless, most people would agree that we

249 Christopher Monckton, "Kicking the Poor in the Teeth: the Cruel Cost of Needless Mitigation," speech at the Heartland Institute, July 2, 2015

shouldn't continue to burn petroleum and natural gas when cleaner, cost effective and sustainable alternatives such as nuclear energy are available.

It has become increasingly obvious that the three "renewables" - wind, solar and biofuels - which are being pushed on us by the climate alarmist cabal - have a very limited application and can never replace carbon based fuels. Trying to force these on people with subsidies, cap-and-trade schemes, regressive taxes and even more draconian tactics will destroy the world's economies and end up killing millions of people. The evidence that these renewables are ineffective substitutes for carbon based fuels is the number of years that these alternatives have been subsidized by government grants - some of which go back to the 1970's. Without such grants these industries would have failed long ago.

Of the amount of the world's energy now being provided by renewables - about 14% - only 0.33% comes from wind and solar. The rest comes from hydroelectric and biofuels such as wood, used mostly in third-world countries.

Economist Bjorn Lomborg has studied this subject extensively. He says that for every dollar governments spend on subsidies, there is a return of only $0.03.[250] So obviously these renewables are not the solution. It is a grievous mistake to force them on people by means of taxes and subsidies.

Since the institution of "green" taxes the cost of heating in Great Britain has risen 63%. During the last five years a million elderly British have been struggling to keep warm, and each year over 2,700 of these people have died from the stress of extended periods of cold weather.[251]

Now there's a call to raise the carbon tax in Ireland from 20 to 80 pounds/ton by 2030. Not only would this be a direct tax, but it would also fuel inflation, the hidden tax.

"There are hundreds of thousands of people who are struggling at the margin of survival...Over 15% of the population has an

---

250 "Economics of climate change innovation | Bjørn Lomborg," TED Talk, Feb. 13, 2017, YouTube Video

251 Ibid.

income below the poverty line...If [an increased] carbon tax comes in, these are the people that are most vulnerable," noted Dr. Sean Healy, head of Social Justice Ireland.[252]

Currently in the US over $100 billion is being paid out yearly in subsidies for renewables. By 2035 that number is expected to rise to about $223 billion annually unless people come to their senses.

Several of the world's greatest economists agree that the money now being wasted could be much better invested in research and development of cost effective energy alternatives that actually work. When these resulting innovations come about, as they ultimately will, they'll quickly be put in place by the free enterprise system. This has happened throughout history. Kerosene replaced whale oil in the 1860's. Automobiles replaced horses in the early 1900's, and the catalytic converter solved the air pollution problem posed by cars. The Green Revolution in agriculture saved millions from starvation in the 1970's.

You simply can't force innovations that don't yet exist. Lomborg estimates that for every dollar invested in research there's been a return of $11. This is certainly a much better return than the $0.03 return for every dollar spent on subsidies.

Wind power is a very old technology which goes back to the 14th century. Before fossil fuels were used, the world's people were trying to get away from renewables such as biofuels (wood) that had been used for thousands of years. Why are these technologies being renewed now?

On the other hand, statistics show the great, almost unlimited potential available from nuclear energy that is awaiting civilization.

Coal will continue to dominate the world's sources of energy until abundant and safe nuclear energy, from thorium or nuclear fusion, becomes commercially available. In contrast, wind turbines work only when the wind is blowing. And they

252 "Carbon Tax To Hit Lowest Earners Hardest, Says Social Justice Ireland,"https://www.thegwpf. om/carbon-tax-to-hit-lowest-earners-hardest-says-social-justice-ireland/, Jan. 4, 2019

must be backed up 5/6 of the way by the combustion of carbon and nuclear based fuels. The same is true of solar voltaic panels, which don't work at night or in cloudy weather.

The greatest strides in technological progress have always come about through the free enterprise system. Experiments in communism and socialism have always ended in disaster. The fate of the Soviet Union and Venezuela are examples of the nonworkability of such political systems. They've ruined the lives of hundreds of millions of people. We need to put a stop to the relentless political push that's been making inroads with young people who don't know or understand history and older people who haven't not learned these lessons from history.

It is a serious mistake to put this kind of power in the hands of corrupt politicians and bureaucrats who have demonstrated ignorance about climate science. Obviously the best thing governments can do right now is to allow the free market to continue to develop these new technologies in the way they always have.

As a voter, you still have some power to help determine the outcome of this situation. Although we are all busy with day-to-day survival, we need to take the time to assess the position of political candidates on this issue and work for the ones that embrace intelligent courses of action. If you are a policy maker yourself, you owe it to yourself and your constituents to get the facts you need to increase your understanding of the issue. This will enable you to not only make the right decisions but also persuade others to do the same.

Appendix 1

Questions for Global Warming Alarmists

1. How is it that in the 1950's, 60's and 70's when the world's carbon dioxide emissions were ***rising*** rapidly, the earth was ***cooling***?

2. During the Medieval Warm Period Earth's average temperature was about two degrees higher than it now is. Humans weren't burning fossil fuels at that time. What caused that temperature increase?

3. What source of energy do you propose to replace the combustion of carbon-based "fossil" fuels and nuclear energy, when it has been established that wind and solar generated energy is undependable and at least three times as expensive?

4. How will solar and wind power help the US economy, when electricity produced by these methods is at least three times as expensive as carbon based and nuclear fuels?

5. How will solar and wind energy power all of the cars, trucks, farm tractors and heavy construction vehicles that are currently used in the US economy?

6. If the global warming hypothesis is valid, why have agencies such as NOAA and NASA found it necessary to change temperature records by lowering earlier temperatures and raising later temperatures?

7. Why did the earth's average temperature increase just as much from 1910 to 1940 when there was much less CO_2 in the atmosphere, as it did during the final two decades of the 1900's?

Appendix 2

Text from the Green New Deal

In reading through the text, please note how much of this plan for the entire reorganization of the US economy from free enterprise to Communism rests on the premise of human-caused global warming which has never been scientifically proved.

H. Res.109 - 116th Congress (2019-2020)

Recognizing the duty of the Federal Government to create a Green New Deal.

Whereas the October 2018 report entitled "Special Report on Global Warming of 1.5 ºC" by the Intergovernmental Panel on Climate Change and the November 2018 Fourth National Climate Assessment report found that—

(1) human activity is the dominant cause of observed climate change over the past century;

(2) a changing climate is causing sea levels to rise and an increase in wildfires, severe storms, droughts, and other extreme weather events that threaten human life, healthy communities, and critical infrastructure;

(3) global warming at or above 2 degrees Celsius beyond preindustrialized levels will cause—

(A) mass migration from the regions most affected by climate change;

(B) more than $500,000,000,000 in lost annual economic output in the United States by the year 2100;

(C) wildfires that, by 2050, will annually burn at least twice as much forest area in the western United States than was typically burned by wildfires in the years preceding 2019;

(D) a loss of more than 99 percent of all coral reefs on Earth;

(E) more than 350,000,000 more people to be exposed globally to deadly heat stress by 2050; and

(F) a risk of damage to $1,000,000,000,000 of public infrastructure and coastal real estate in the United States; and

(4) global temperatures must be kept below 1.5 degrees Celsius above preindustrialized levels to avoid the most severe impacts of a changing climate, which will require—

(A) global reductions in greenhouse gas emissions from human sources of 40 to 60 percent from 2010 levels by 2030; and

(B) net-zero global emissions by 2050;

Whereas, because the United States has historically been responsible for a disproportionate amount of greenhouse gas emissions, having emitted 20 percent of global greenhouse gas emissions through 2014, and has a high technological capacity, the United States must take a leading role in reducing emissions through economic transformation;

Whereas the United States is currently experiencing several related crises, with—

(1) life expectancy declining while basic needs, such as clean air, clean water, healthy food, and adequate health care, housing, transportation, and education, are inaccessible to a significant portion of the United States population;

(2) a 4-decade trend of wage stagnation, deindustrialization, and antilabor policies that has led to—

(A) hourly wages overall stagnating since the 1970s despite increased worker productivity;

(B) the third-worst level of socioeconomic mobility in the developed world before the Great Recession;

(C) the erosion of the earning and bargaining power of workers in the United States; and

(D) inadequate resources for public sector workers to confront the challenges of climate change at local, State, and Federal levels; and

(3) the greatest income inequality since the 1920s, with —

(A) the top 1 percent of earners accruing 91 percent of gains in the first few years of economic recovery after the Great Recession;

(B) a large racial wealth divide amounting to a difference of 20 times more wealth between the average white family and the average black family; and

(C) a gender earnings gap that results in women earning approximately 80 percent as much as men, at the median;

Whereas climate change, pollution, and environmental destruction have exacerbated systemic racial, regional, social, environmental, and economic injustices (referred to in this preamble as "systemic injustices") by disproportionately affecting indigenous peoples, communities of color, migrant communities, deindustrialized communities, depopulated rural communities, the poor, low-income workers, women, the elderly, the unhoused, people with disabilities, and youth (referred to in this preamble as "frontline and vulnerable communities");

Whereas, climate change constitutes a direct threat to the national security of the United States—

(1) by impacting the economic, environmental, and social stability of countries and communities around the world; and

(2) by acting as a threat multiplier;

Whereas the Federal Government-led mobilizations during World War II and the New Deal created the greatest middle class that the United States has ever seen, but many members of frontline and vulnerable communities were excluded from many of the economic and societal benefits of those mobilizations; and

Whereas the House of Representatives recognizes that a new national, social, industrial, and economic mobilization on a scale not seen since World War II and the New Deal era is a historic opportunity—

(1) to create millions of good, high-wage jobs in the United States;

(2) to provide unprecedented levels of prosperity and economic security for all people of the United States; and

(3) to counteract systemic injustices: Now, therefore, be it Resolved, That it is the sense of the House of Representatives that—

(1) it is the duty of the Federal Government to create a Green New Deal—

(A) to achieve net-zero greenhouse gas emissions through a fair and just transition for all communities and workers;

(B) to create millions of good, high-wage jobs and ensure

prosperity and economic security for all people of the United States;

(C) to invest in the infrastructure and industry of the United States to sustainably meet the challenges of the 21st century;

(D) to secure for all people of the United States for generations to come—

(i) clean air and water;

(ii) climate and community resiliency;

(iii) healthy food;

(iv) access to nature; and

(v) a sustainable environment; and

(E) to promote justice and equity by stopping current, preventing future, and repairing historic oppression of indigenous peoples, communities of color, migrant communities, deindustrialized communities, depopulated rural communities, the poor, low-income workers, women, the elderly, the unhoused, people with disabilities, and youth (referred to in this resolution as "frontline and vulnerable communities");

(2) the goals described in subparagraphs (A) through (E) of paragraph (1) (referred to in this resolution as the "Green New Deal goals") should be accomplished through a 10-year national mobilization (referred to in this resolution as the "Green New Deal mobilization") that will require the following goals and projects—

(A) building resiliency against climate change-related disasters, such as extreme weather, including by leveraging funding and providing investments for community-defined projects and strategies;

(B) repairing and upgrading the infrastructure in the United States, including—

(i) by eliminating pollution and greenhouse gas emissions as much as technologically feasible;

(ii) by guaranteeing universal access to clean water;

(iii) by reducing the risks posed by climate impacts; and

(iv) by ensuring that any infrastructure bill

considered by Congress addresses climate change;

(C) meeting 100 percent of the power demand in the United States through clean, renewable, and zero-emission energy sources, including—

(i) dramatically expanding and upgrading renewable power sources; and

(ii) by deploying new capacity;

(D) building or upgrading to energy-efficient, distributed, and "smart" power grids, and ensuring affordable access to electricity;

(E) upgrading all existing buildings in the United States and building new buildings to achieve maximum energy efficiency, water efficiency, safety, affordability, comfort, and durability, including through electrification;

(F) spurring massive growth in clean manufacturing in the United States and removing pollution and greenhouse gas emissions from manufacturing and industry as much as is technologically feasible, including by expanding renewable energy manufacturing and investing in existing manufacturing and industry;

(G) working collaboratively with farmers and ranchers in the United States to remove pollution and greenhouse gas emissions from the agricultural sector as much as is technologically feasible, including—

(i) by supporting family farming;

(ii) by investing in sustainable farming and land use practices that increase soil health; and

(iii) by building a more sustainable food system that ensures universal access to healthy food;

(H) overhauling transportation systems in the United States to remove pollution and greenhouse gas emissions from the transportation sector as much as is technologically feasible, including through investment in—

(i) zero-emission vehicle infrastructure and manufacturing;

(ii) clean, affordable, and accessible public transit; and

(iii) high-speed rail;

(I) mitigating and managing the long-term adverse health, economic, and other effects of pollution and climate change, including by providing funding for community-defined projects and strategies;

(J) removing greenhouse gases from the atmosphere and reducing pollution by restoring natural ecosystems through proven low-tech solutions that increase soil carbon storage, such as land preservation and afforestation;

(K) restoring and protecting threatened, endangered, and fragile ecosystems through locally appropriate and science-based projects that enhance biodiversity and support climate resiliency;

(L) cleaning up existing hazardous waste and abandoned sites, ensuring economic development and sustainability on those sites;

(M) identifying other emission and pollution sources and creating solutions to remove them; and

(N) promoting the international exchange of technology, expertise, products, funding, and services, with the aim of making the United States the international leader on climate action, and to help other countries achieve a Green New Deal;

(3) a Green New Deal must be developed through transparent and inclusive consultation, collaboration, and partnership with frontline and vulnerable communities, labor unions, worker cooperatives, civil society groups, academia, and businesses; and

(4) to achieve the Green New Deal goals and mobilization, a Green New Deal will require the following goals and projects—

(A) providing and leveraging, in a way that ensures that the public receives appropriate ownership stakes and returns on investment, adequate capital (including through community grants, public banks, and other public financing), technical expertise, supporting policies, and other forms of assistance to communities, organizations, Federal, State, and local government agencies, and businesses working on the Green New Deal mobilization;

(B) ensuring that the Federal Government takes into account the complete environmental and social costs and impacts of emissions through—

(i) existing laws;

(ii) new policies and programs; and

(iii) ensuring that frontline and vulnerable communities shall not be adversely affected;

(C) providing resources, training, and high-quality education, including higher education, to all people of the United States, with a focus on frontline and vulnerable communities, so that all people of the United States may be full and equal participants in the Green New Deal mobilization;

(D) making public investments in the research and development of new clean and renewable energy technologies and industries;

(E) directing investments to spur economic development, deepen and diversify industry and business in local and regional economies, and build wealth and community ownership, while prioritizing high-quality job creation and economic, social, and environmental benefits in frontline and vulnerable communities, and deindustrialized communities, that may otherwise struggle with the transition away from greenhouse gas intensive industries;

(F) ensuring the use of democratic and participatory processes that are inclusive of and led by frontline and vulnerable communities and workers to plan, implement, and administer the Green New Deal mobilization at the local level;

(G) ensuring that the Green New Deal mobilization creates high-quality union jobs that pay prevailing wages, hires local workers, offers training and advancement opportunities, and guarantees wage and benefit parity for workers affected by the transition;

(H) guaranteeing a job with a family-sustaining wage, adequate family and medical leave, paid vacations, and retirement security to all people of the United States;

(I) strengthening and protecting the right of all workers to organize, unionize, and collectively bargain free of coercion, intimidation, and harassment;

(J) strengthening and enforcing labor, workplace health and safety, antidiscrimination, and wage and hour standards across all employers, industries, and sectors;

(K) enacting and enforcing trade rules, procurement standards, and border adjustments with strong labor and environmental protections—

(i) to stop the transfer of jobs and pollution overseas; and

(ii) to grow domestic manufacturing in the U.S.

(L) ensuring that public lands, waters, and oceans are protected and that eminent domain is not abused;

(M) obtaining the free, prior, and informed consent of indigenous peoples for all decisions that affect indigenous peoples and their traditional territories, honoring all treaties and agreements with indigenous peoples, and protecting and enforcing the sovereignty and land rights of indigenous peoples;

(N) ensuring a commercial environment where every businessperson is free from unfair competition and domination by domestic or international monopolies; and

(O) providing all people of the United States with—

(i) high-quality health care;

(ii) affordable, safe, and adequate housing;

(iii) economic security; and

(iv) clean water, clean air, healthy and affordable food, and access to nature.

Appendix 3

Glossary of Terms and Acronyms

AGW anthropogenic, or human-caused, global warming

albedo effect the fraction of the total light striking a surface that is reflected, depending on the brightness of the surface. Clouds and ice have a high albedo, because they reflect sunlight.

biofuel fuel (such as wood or ethanol) composed of or produced from biological raw materials.

CCUS carbon capture and storage

climate weather, usually in a certain area, over an extended period of time, 20 years or more

climate denier derogatory term given to anyone who disagrees with the official IPCC agenda, even if is a legitimate criticism of the science. This is a logical fallacy known as an "ad hominem" (Latin: "against the man") attack.

COP Conference of the Parties. The United Nations Climate Change Conferences are yearly conferences held in the framework of the United Nations Framework Convention on Climate Change (UNFCCC). They serve as the formal meeting of the UNFCCC Parties (Conference of the Parties, COP) to assess progress in dealing with climate change, and beginning in the mid-1990s, to negotiate the Kyoto Protocol to establish legally binding obligations for developed countries to reduce their greenhouse gas emissions.[253]

cosmic rays highly energetic particles moving at nearly the speed of light that enter Earth's atmosphere

CRU Climate Research Unit, University of East Anglia in Britain

---

253 From Wikipedia, "The Climate Change Conference"

data information obtained from direct observation and measurement

EPA Environmental Protection Administration

fission process of splitting the nucleus of an atom of a radioactive element to produce smaller atoms, releasing a great amount of energy

fusion process of forcing atoms of lighter elements to join together, using extreme heat and pressure, resulting in the release of a tremendous amount of energy

GISS Goddard Institute for Space Studies, part of NASA

heat exchanger a device that is used to transfer heat from one fluid (liquid or gas) to another without allowing them to mix

hypothesis an idea or explanation that has not yet been verified by experiment or observation

IPCC Intergovernmental Panel on Climate Change, a United Nations agency

isostasy maintaining equilibrium in the earth's crust such that the forces tending to elevate landmasses balance the forces tending to depress landmasses.

Kyoto Protocol (1997) an international treaty extending the 1992 United Nations Framework Convention on Climate Change It commits state parties to reduce greenhouse gas emissions

mean average

NCAR National Center for Atmospheric Research, a US government agency involved in climate models

NCDC National Climactic Data Center

net-zero carbon a method of energy generation in which the carbon dioxide emitted in producing energy is balanced by carbon sinks that are specifically created to offset it. Carbon sinks (+), such as forests, and carbon emissions (-) are added together to reach zero.

NGO nongovernmental organization: An organized, highly financed group that tries to achieve social or political aims but supposedly is not controlled by a government

NOAA National Oceanic and Atmospheric Administration, or "National Weather Service", part of the US Dept. of Commerce

peer review evaluation of a scientist's work by qualified people working in the same scientific discipline

proxy data climate data that is derived from clues other than weather instruments, such as ice core samples, tree rings, etc.

RED [European Union] Renewable Energy Directives

scientism an exaggerated, unrealistic trust in the ability of natural science to solve all of our problems

solar wind a stream of protons originating on the sun that, when arriving at Earth, help deflect cosmic rays from outer space, thus preventing the formation of low clouds that block solar radiation and cool the climate

TSI total solar irradiance

UNEP United Nations Environment Program

UNFCCC United Nations Framework Convention on Climate Change

USHCN United States Historical Climatology Network

WMO World Meteorological Organization

Made in the USA
Monee, IL
04 November 2026

67385895R00105